STUDENT SOLUTIONS MANUAL FOR

Finite Mathematics

FOR THE MANAGERIAL, LIFE, AND SOCIAL SCIENCES

SIXTH EDITION

S. T. TAN
Stonehill College

Brooks/Cole
Thomson Learning

Australia • Canada • Mexico • Singapore • Spain
United Kingdom • United States

Assistant Editor: *Stephanie Schmidt*
Marketing Team: *Karin Sandberg, Beth Kroenke*
Editorial Assistant: *Emily Davidson*
Production Coordinator: *Dorothy Bell*

Cover Design: *Lisa Henry*
Cover Illustration: *Judith Harkness*
Print Buyer: *Micky Lawler*
Printing and Binding: *Webcom Limited*

For more information about this or any other Brooks/Cole product, contact:
BROOKS/COLE
511 Forest Lodge Road
Pacific Grove, CA 93950 USA
www.brookscole.com
1-800-423-0563 (Thomson Learning Academic Resource Center)

For permission to use material from this work, contact us by
Web: www.thomsonrights.com
fax: 1-800-730-2215
phone: 1-800-730-2214

Printed in Canada

10 9 8 7 6 5 4 3 2 1

ISBN 0-534-37001-2

FINITE MATHEMATICS
For The Managerial, Life, and Social Sciences

CONTENTS

CHAPTER 6   SETS AND COUNTING

CHAPTER 7   PROBABILITY

CHAPTER 8   PROBABILITY DISTRIBUTIONS AND STATISTICS

CHAPTER 9   MARKOV CHAINS AND THE THEORY OF GAMES

CHAPTER 1

EXERCISES 1.1, page 8

1. The coordinates of A are (3,3) and it is located in Quadrant I.

3. The coordinates of C are (2,-2) and it is located in Quadrant IV.

5. The coordinates of E are (-4,-6) and it is located in Quadrant III.

7. A   9. E, F, and G.   11. F

For Exercises 13-19, refer to the following figure.

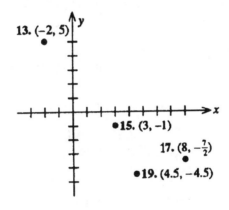

21. Using the distance formula, we find that $\sqrt{(4-1)^2+(7-3)^2}=\sqrt{3^2+4^2}=\sqrt{25}=5$.

23. Using the distance formula, we find that
$$\sqrt{(4-(-1))^2+(9-3)^2}=\sqrt{5^2+6^2}=\sqrt{25+36}=\sqrt{61}.$$

25. The coordinates of the points have the form $(x,-6)$. Since the points are 10 units away from the origin, we have
$$(x-0)^2+(-6-0)^2=10^2$$
$$x^2=64,$$
or $x=\pm 8$. Therefore, the required points are $(-8,-6)$ and $(8,-6)$.

27. The points are shown in the diagram that follows.

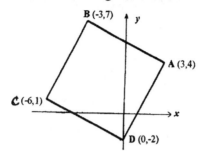

To show that the four sides are equal, we compute the following:

$$d(A,B) = \sqrt{(-3-3)^2 + (7-4)^2} = \sqrt{(-6)^2 + 3^2} = \sqrt{45}$$
$$d(B,C) = \sqrt{[(-6-(-3)]^2 + (1-7)^2} = \sqrt{(-3)^2 + (-6)^2} = \sqrt{45}$$
$$d(C,D) = \sqrt{[0-(-6)]^2 + [(-2)-1]^2} = \sqrt{(6)^2 + (-3)^2} = \sqrt{45}$$
$$d(A,D) = \sqrt{(0-3)^2 + (-2-4)^2} = \sqrt{(3)^2 + (-6)^2} = \sqrt{45}.$$

Next, to show that $\triangle ABC$ is a right triangle, we show that it satisfies the Pythagorean Theorem. Thus,

$$d(A,C) = \sqrt{(-6-3)^2 + (1-4)^2} = \sqrt{(-9)^2 + (-3)^2} = \sqrt{90} = 3\sqrt{10}$$

and $[d(A,B)]^2 + [d(B,C)]^2 = 90 = [d(A,C)]^2$. Similarly, $d(B,D) = \sqrt{90} = 3\sqrt{10}$, so $\triangle BAD$ is a right triangle as well. It follows that $\angle B$ and $\angle D$ are right angles, and we conclude that $ADCB$ is a square

29. The equation of the circle with radius 5 and center (2,-3) is given by
$$(x-2)^2 + [y-(-3)]^2 = 5^2$$
or $$(x-2)^2 + (y+3)^2 = 25.$$

31. The equation of the circle with radius 5 and center (0, 0) is given by
$$(x-0)^2 + (y-0)^2 = 5^2$$
or $$x^2 + y^2 = 25$$

33. The distance between the points (5,2) and (2,-3) is given by
$$d = \sqrt{(5-2)^2 + (2-(-3))^2} = \sqrt{3^2 + 5^2} = \sqrt{34}.$$

Therefore $r = \sqrt{34}$ and the equation of the circle passing through (5,2) and (2,-3) is

$$(x-2)^2 + [y-(-3)]^2 = 34$$

or $\quad (x-2)^2 + (y+3)^2 = 34.$

35. Referring to the diagram on page 12 of the text, we see that the distance from A to B is given by $d(A,B) = \sqrt{400^2 + 300^2} = \sqrt{250,000} = 500.$ The distance from B to C is given by

$$d(B,C) = \sqrt{(-800-400)^2 + (800-300)^2} = \sqrt{(-1200)^2 + (500)^2}$$
$$= \sqrt{1,690,000} = 1300.$$

The distance from C to D is given by

$$d(C,D) = \sqrt{[-800-(-800)]^2 + (800-0)^2} = \sqrt{0+800^2} = 800 .$$

The distance from D to A is given by

$$d(D,A) = \sqrt{[0-(-800)]^2 + (0-0)} = \sqrt{640000} = 800.$$

Therefore, the total distance covered on the tour, is

$$d(A,B) + d(B,C) + d(C,D) + d(D,A) = 500 + 1300 + 800 + 800$$
$$= 3400, \quad \text{or } 3400 \text{ miles.}$$

37. Referring to the following diagram,

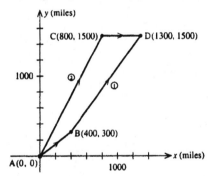

we see that the distance he would cover if he took Route (1) is given by

$$d(A,B) + d(B,D) = \sqrt{400^2 + 300^2} + \sqrt{(1300-400)^2 + (1500-300)^2}$$
$$= \sqrt{250,000} + \sqrt{2,250,000} = 500 + 1500 = 2000,$$

or 2000 miles. On the other hand, the distance he would cover if he took Route (2) is given by

$$d(A,C) + d(C,D) = \sqrt{800^2 + 1500^2} + \sqrt{(1300-800)^2}$$
$$= \sqrt{2,890,000} + \sqrt{250,000} = 1700 + 500 = 2200,$$

or 2200 miles. Comparing these results, we see that he should take Route (1).

39. Calculations to determine VHF requirements:

$$d = \sqrt{25^2 + 35^2} = \sqrt{625 + 1225} = \sqrt{1850} \approx 43.01.$$

Models B through D satisfy this requirement.

Calculations to determine UHF requirements:

$$d = \sqrt{20^2 + 32^2} = \sqrt{400 + 1024} = \sqrt{1424} = 37.74$$

Models C through D satisfy this requirement. Therefore, Model C will allow him to receive both channels at the least cost.

41. a. Let the position of ship A and Ship B after t hours be $A(0,y)$ and $B(x, y)$, respectively. Then $x = 30t$ and $y = 20t$. Therefore, the distance between the two ships is

$$D = \sqrt{(30t)^2 + (20t)^2} = \sqrt{900t^2 + 400t^2} = 10\sqrt{13}t.$$

b. The required distance is obtained by letting $t = 2$ giving $D = 10\sqrt{13}(2)$ or approximately 72.11 miles.

43. True. $kx^2 + ky^2 = a^2$; $x^2 + y^2 = \dfrac{a^2}{k} < a^2$ if $k > 1$. So the radius of the circle with equation (1) is a circle of radius smaller than a if $k > 1$ (and centered at the origin). Therefore, it lies inside the circle of radius a with equation $x^2 + y^2 = a^2$.

45. Referring to the figure in the text, we see that the distance between the two points is given by the length of the hypotenuse of the right triangle. That is,

$$d = \sqrt{(x_2 - x_1)^2 + (y_2 - y_1)^2}$$

EXERCISES 1.2, page 25

1. Referring to the figure shown in the text, we see that $m = \dfrac{2 - 0}{0 - (-4)} = \dfrac{1}{2}$.

3. This is a vertical line, and hence its slope is undefined.

5. $m = \dfrac{y_2 - y_1}{x_2 - x_1} = \dfrac{8 - 3}{5 - 4} = 5.$

7. $m = \dfrac{y_2 - y_1}{x_2 - x_1} = \dfrac{8 - 3}{4 - (-2)} = \dfrac{5}{6}.$

9. $m = \dfrac{y_2 - y_1}{x_2 - x_1} = \dfrac{d - b}{c - a}.$

11. Since the equation is in the slope-intercept form, we read off the slope $m = 4$.
    a. If x increases by 1 unit, then y increases by 4 units.
    b. If x decreases by 2 units, y decreases by $4(-2) = -8$ units.

13. The slope of the line through A and B is $\dfrac{-10-(-2)}{-3-1} = \dfrac{-8}{-4} = 2$.

The slope of the line through C and D is $\dfrac{1-5}{-1-1} = \dfrac{-4}{-2} = 2$.

Since the slopes of these two lines are equal, the lines are parallel.

15. The slope of the line through A and B is $\dfrac{2-5}{4-(-2)} = -\dfrac{3}{6} = -\dfrac{1}{2}$.

The slope of the line through C and D is $\dfrac{6-(-2)}{3-(-1)} = \dfrac{8}{4} = 2$.

Since the slopes of these two lines are the negative reciprocals of each other, the lines are perpendicular.

17. The slope of the line through the point $(1, a)$ and $(4,- 2)$ is $m_1 = \dfrac{-2-a}{4-1}$ and the slope of the line through $(2,8)$ and $(-7, a + 4)$ is $m_2 = \dfrac{a+4-8}{-7-2}$. Since these two lines are parallel, m_1 is equal to m_2. Therefore,

$$\frac{-2-a}{3} = \frac{a-4}{-9}$$
$$-9(-2-a) = 3(a-4)$$
$$18+9a = 3a-12$$
$$6a = -30 \qquad \text{and} \quad a = -5$$

19. An equation of a horizontal line is of the form $y = b$. In this case $b = -3$, so $y = -3$ is an equation of the line.

21. e 23. a 25. f

27. We use the point-slope form of an equation of a line with the point $(3,- 4)$ and slope $m = 2$. Thus

$$y - y_1 = m(x - x_1),$$
and
$$y - (- 4) = 2(x - 3)$$
$$y + 4 = 2x - 6$$
$$y = 2x - 10.$$

29. Since the slope $m = 0$, we know that the line is a horizontal line of the form $y = b$. Since the line passes through $(-3,2)$, we see that $b = 2$, and an equation of the line is $y = 2$.

31. We first compute the slope of the line joining the points (2,4) and (3,7). Thus,
$$m = \frac{7-4}{3-2} = 3.$$
Using the point-slope form of an equation of a line with the point (2,4) and slope $m = 3$, we find
$$y - 4 = 3(x - 2)$$
$$y = 3x - 2.$$

33. We first compute the slope of the line joining the points (1,2) and (−3,−2). Thus,
$$m = \frac{-2-2}{-3-1} = \frac{-4}{-4} = 1.$$
Using the point-slope form of an equation of a line with the point (1,2) and slope $m = 1$, we find
$$y - 2 = x - 1$$
$$y = x + 1.$$

35. We use the slope-intercept form of an equation of a line: $y = mx + b$. Since $m = 3$, and $b = 4$, the equation is $y = 3x + 4$.

37. We use the slope-intercept form of an equation of a line: $y = mx + b$. Since $m = 0$, and $b = 5$, the equation is $y = 5$.

39. We first write the given equation in the slope-intercept form:
$$x - 2y = 0$$
$$-2y = -x$$
$$y = \tfrac{1}{2}x .$$
From this equation, we see that $m = 1/2$ and $b = 0$.

41. We write the equation in slope-intercept form:
$$2x - 3y - 9 = 0$$
$$-3y = -2x + 9$$
$$y = \tfrac{2}{3}x - 3.$$
From this equation, we see that $m = 2/3$ and $b = -3$.

43. We write the equation in slope-intercept form:
$$2x + 4y = 14$$
$$4y = -2x + 14$$
$$y = -\tfrac{2}{4}x + \tfrac{14}{4}$$
$$= -\tfrac{1}{2}x + \tfrac{7}{2}.$$

From this equation, we see that $m = -1/2$ and $b = 7/2$.

45. We first write the equation $2x - 4y - 8 = 0$ in slope- intercept form:
$$2x - 4y - 8 = 0$$
$$4y = 2x - 8$$
$$y = \tfrac{1}{2}x - 2$$
Now the required line is parallel to this line, and hence has the same slope. Using the point-slope equation of a line with $m = 1/2$ and the point $(-2,2)$, we have
$$y - 2 = \tfrac{1}{2}[x - (-2)]$$
$$y = \tfrac{1}{2}x + 3.$$

47. A line parallel to the x-axis has slope 0 and is of the form $y = b$. Since the line is 6 units below the axis, it passes through $(0,-6)$ and its equation is $y = -6$.

49. We use the point-slope form of an equation of a line to obtain
$$y - b = 0(x - a) \quad \text{or} \quad y = b.$$

51. Since the required line is parallel to the line joining $(-3,2)$ and $(6,8)$, it has slope
$$m = \frac{8 - 2}{6 - (-3)} = \frac{6}{9} = \frac{2}{3}.$$
We also know that the required line passes through $(-5,-4)$. Using the point-slope form of an equation of a line, we find
$$y - (-4) = \frac{2}{3}(x - (-5))$$
or
$$y = \tfrac{2}{3}x + \tfrac{10}{3} - 4$$
that is
$$y = \tfrac{2}{3}x - \tfrac{2}{3} \quad .$$

53. Since the point $(-3,5)$ lies on the line $kx + 3y + 9 = 0$, it satisfies the equation. Substituting $x = -3$ and $y = 5$ into the equation gives
$$-3k + 15 + 9 = 0$$
or
$$k = 8.$$

55. $3x - 2y + 6 = 0$

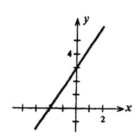

57. $x + 2y - 4 = 0$

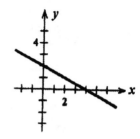

59. $y + 5 = 0$

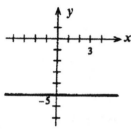

61. Since the line passes through the points $(a, 0)$ and $(0, b)$, its slope is $m = \dfrac{b-0}{0-a} = -\dfrac{b}{a}$.

Then, using the point-slope form of an equation of a line with the point $(a, 0)$ we

have
$$y - 0 = -\tfrac{b}{a}(x - a)$$
$$y = -\tfrac{b}{a}x + b$$
which may be written in the form
$$\tfrac{b}{a}x + y = b.$$
Multiplying this last equation by $1/b$, we have
$$\frac{x}{a} + \frac{y}{b} = 1.$$

63. Using the equation $\dfrac{x}{a} + \dfrac{y}{b} = 1$ with $a = -2$ and $b = -4$, we have $-\dfrac{x}{2} - \dfrac{y}{4} = 1$.

Then
$$-4x - 2y = 8$$
$$2y = -8 - 4x$$
$$y = -2x - 4.$$

65. Using the equation $\frac{x}{a}+\frac{y}{b}=1$ with $a=4$ and $b=-1/2$, we have

$$\frac{x}{4}+\frac{y}{-\frac{1}{2}}=1$$

$$-\tfrac{1}{4}x+2y=-1$$

$$2y=\tfrac{1}{4}x-1$$

$$y=\tfrac{1}{8}x-\tfrac{1}{2}.$$

67. The slope of the line passing through A and B is $m=\dfrac{7-1}{1-(-2)}=\dfrac{6}{3}=2$,

and the slope of the line passing through B and C is $m=\dfrac{13-7}{4-1}=\dfrac{6}{3}=2$.

Since the slopes are equal, the points lie on the same line.

69. a. $y=0.55x$

b. Solving the equation $1100=0.55x$ for x, we have $x=\dfrac{1100}{0.55}=2000$.

71. Using the points $(0, 0.68)$ and $(10, 0.80)$, we see that the slope of the required line is

$$m=\frac{0.80-0.68}{10-0}=\frac{0.12}{10}=.012.$$

Next, using the point-slope form of the equation of a line, we have

$$y-0.68=0.012(t-0)$$

or $\qquad\qquad y=0.012t+0.68.$

Therefore, when $t=12$, we have

$$y=0.012(12)+0.68=.824$$

or 82.4%. That is, in 2002 women's wages are expected to be 82.4% of men's wages.

73. a. – b.

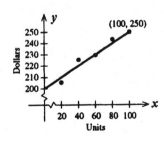

c. Using the points (0,200) and (100,250), we see that the slope of the required line is $m = \dfrac{250-200}{100} = \dfrac{1}{2}$. Therefore, the required equation is
$$y - 200 = \tfrac{1}{2}x \quad \text{or} \quad y = \tfrac{1}{2}x + 200.$$

d. The approximate cost for producing 54 units of the commodity is
$\tfrac{1}{2}(54) + 200,$ or $227.

75   a. – b.

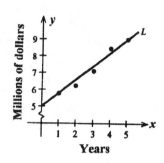

c. The slope of L is $m = \dfrac{9.0-5.8}{5-1} = \dfrac{3.2}{4} = 0.8$. Using the point-slope form of an equation of a line, we have $y - 5.8 = 0.8(x-1) = 0.8x - 0.8$, or $y = 0.8x + 5$.

d. Using the equation of part (c) with $x = 9$, we have
$$y = 0.8(9) + 5 = 12.2, \quad \text{or } 12.2 \text{ million.}$$

77.   a. We obtain a family of parallel lines each having slope m.
b. We obtain a family of straight lines all of which pass through the point $(0,b)$.

79.   True. The slope of the line $Ax + By + C = 0$ is $-B/A$. (Write it in the slope-intercept form.) Similarly, the slope of the line $ax + by + c = 0$ is $-b/a$. They are parallel if and only if $-\dfrac{B}{A} = -\dfrac{b}{a}$, $Ab = aB$, or $Ab - aB = 0$.

81.   True. The slope of the line $ax + by + c_1 = 0$ is $m_1 = -a/b$. The slope of the line $bx - ay + c_2 = 0$ is $m_2 = b/a$. Since $m_1 m_2 = -1$, the straight lines are indeed perpendicular.

83.   Writing each equation in the slope-intercept form, we have
$$y = -\frac{a_1}{b_1}x - \frac{c_1}{b_1} \quad (b_1 \neq 0) \quad \text{and} \quad y = -\frac{a_2}{b_2}x - \frac{c_2}{b_2} \quad (b_2 \neq 0)$$

Since two lines are parallel if and only if their slopes are equal, we see that the lines are parallel if and only if $-\dfrac{a_1}{b_1} = -\dfrac{a_2}{b_2}$, or $a_1b_2 - b_1a_2 = 0$.

USING TECHNOLOGY EXERCISES 1.2, page 31

1.

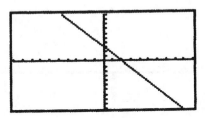

3.

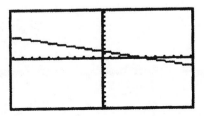

5.

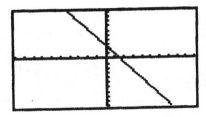

7. a.

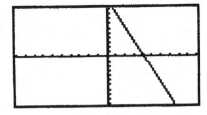

b.

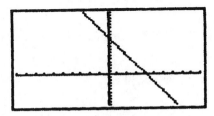

9. a.

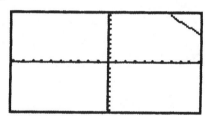

b.

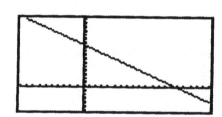

11.

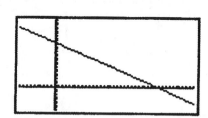

13.

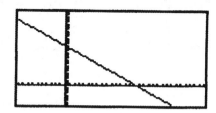

15.

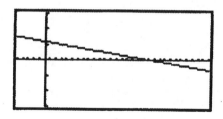

17.

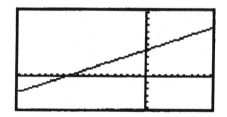

EXERCISES 1.3, page 41

1. Yes. Solving for y in terms of x, we find $3y = -2x + 6$, or $y = -\frac{2}{3}x + 2$.

3. Yes. Solving for y in terms of x, we find $2y = x + 4$, or $y = \frac{1}{2}x + 2$.

5. Yes. Solving for y in terms of x, we have $4y = 2x + 9$, or $y = \frac{1}{2}x + \frac{9}{4}$.

7. y is not a linear function of x because of the quadratic term $2x^2$.

9. y is not a linear function of x because of the term $-3y^2$.

11. a. $C(x) = 8x + 40,000$, where x is the number of units produced.
    b. $R(x) = 12x$, where x is the number of units sold.
    c. $P(x) = R(x) - C(x) = 12x - (8x + 40,000) = 4x - 40,000$.
    d. $P(8,000) = 4(8,000) - 40,000 = -8,000$, or a loss of \$8,000.
       $P(12,000) = 4(12,000) - 40,000 = 8,000$ or a profit of \$8,000.

13. $f(0) = 2$ gives $m(0) + b = 2$, or $b = 2$. So, $f(x) = mx + 2$. Next, $f(3) = -1$ gives $m(3) + 2 = -1$, or $m = -1$.

15. Let V be the book value of the office building after 2000. Since $V = 1,000,000$ when $t = 0$, the line passes through $(0, 1,000,000)$. Similarly, when $t = 50$, $V = 0$, so the line passes through $(50, 0)$. Then the slope of the line is given by
$$m = \frac{0 - 1,000,000}{50 - 0} = -20,000.$$
Using the point-slope form of the equation of a line with the point $(0, 1,000,000)$, we have $\quad V - 1,000,000 = -20,000(t - 0)$,
or $\qquad\qquad\qquad V = -20,000t + 1,000,000$.
In 2005, $t = 5$ and $V = -20,000(5) + 1,000,000 = 900,000$, $\quad$ or \$900,000.
In 2010, $t = 10$ and $V = -20,000(10) + 1,000,000 = 800,000$, $\quad$ or \$800,000.

17. The consumption function is given by $C(x) = 0.75x + 6$. When $x = 0$, we have $C(0) = 0.75(0) + 6 = 6$, or \$6 billion dollars.
When $x = 50$, $\qquad C(50) = 0.75(50) + 6 = 43.5$, $\quad$ or \$43.5 billion dollars.
When $x = 100$, $\quad C(100) = 0.75(100) + 6 = 81$, $\quad$ or \$81 billion dollars.

19. a. $y = 1.053x$, where x is the monthly benefit before adjustment, and y is the adjusted monthly benefit.
    b. His adjusted monthly benefit will be $(1.053)(620) = 652.86$, $\quad$ or \$652.86.

21. Let the number of tapes produced and sold be x. Then
$$C(x) = 12,100 + 0.60x$$
$$R(x) = 1.15x$$
and $\qquad P(x) = R(x) - C(x) = 1.15x - (12,100 + 0.60x)$
$$= 0.55x - 12,100.$$

23. Let the value of the minicomputer after t years be V. When $t = 0$, $V = 60,000$ and when $t = 4$, $V = 12,000$.

a. Since $\quad m = \dfrac{12{,}000 - 60{,}000}{4} = -\dfrac{48{,}000}{4} = -12{,}000$

the rate of depreciation $(-m)$ is \$12,000/yr.

b. Using the point-slope form of the equation of a line with the point (4, 12,000), we have $\quad V - 12{,}000 = -12{,}000(t - 4)$

or $\quad\quad\quad\quad\quad V = -12{,}000t + 60{,}000.$

c.

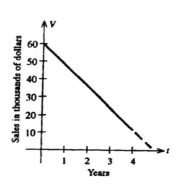

d. When $t = 3$, $V = -12{,}000(3) + 60{,}000 = 24{,}000,$ or \$24,000.

25. The formula given in Exercise 24 is $V = C - \dfrac{C - S}{N} t$.

Then, when $C = 1{,}000{,}000$, $N = 50$, and $S = 0$, we have

$$V = 1{,}000{,}000 - \dfrac{1{,}000{,}000 - 0}{50} t \quad \text{or} \quad V = 1{,}000{,}000 - 20{,}000t.$$

In 2005, $t = 5$ and $\quad V = 1{,}000{,}000 - 20{,}000(5) = 900{,}000,$ or \$900,000.

In 2010, $t = 10$ and $\quad V = 1{,}000{,}000 - 20{,}000(10) = 800{,}000$ or \$800,000.

27. a. $D(S) = \dfrac{Sa}{1.7}$. If we think of D as having the form $D(S) = mS + b$, then

$m = \dfrac{a}{1.7}$, $\quad b = 0$, and D is a linear function of S.

b. $D(0.4) = \dfrac{500(0.4)}{1.7} = 117.647,$ or approximately 117.65 mg.

29. a. Since the relationship is linear, we can write $F = mC + b$, where m and b are constants. Using the condition $C = 0$ when $F = 32$, we have $32 = b$, and so $F = mC + 32$. Next, using the condition $C = 100$ when $F = 212$, we have

$$212 = 100m + 32 \quad \text{or} \quad m = \tfrac{9}{5}.$$

Therefore, $F = \tfrac{9}{5}C + 32$.

b. From (a), we see $F = \tfrac{9}{5}C + 32$. Next, when $C = 20$,

$$F = \tfrac{9}{5}(20) + 32 = 68$$

and so the temperature equivalent to 20°C is 68°F.

c. Solving for C in terms of F, we find $\tfrac{9}{5}C = F - 32$, or $C = \tfrac{5}{9}F - \tfrac{160}{9}$.
When $F = 70$, $C = \tfrac{5}{9}(70) - \tfrac{160}{9} = \tfrac{190}{9}$, or approximately 21.1°C.

31. The slope of L_2 is greater than that of L_1. This tells us that if the manufacturer lowers the unit price for each model clock radio by the same amount, the quantity demanded of model B radios will be greater than that of the model A radios.

33. a. Setting $x = 0$, gives $3p = 18$, or $p = 6$. Next, setting $p = 0$, gives $2x = 18$, or $x = 9$.

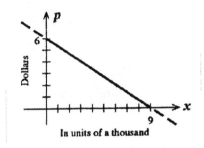

b. When $p = 4$,
$$2x + 3(4) - 18 = 0$$
$$2x = 18 - 12 = 6$$
and $x = 3$. Therefore, the quantity demanded when $p = 4$ is 3000. (Remember x is given in units of a thousand.)

35. a. When $x = 0$, $p = 60$ and when $p = 0$, $-3x = -60$, or $x = 20$.

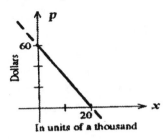

b. When $p = 30$, $\qquad 30 = -3x + 60$

$\qquad\qquad\qquad\qquad 3x = 30 \qquad\qquad$ and $\quad x = 10$.

Therefore, the quantity demanded when $p = 30$ is 10,000 units.

37. When $x = 1000$, $p = 55$, and when $x = 600$, $p = 85$. Therefore, the graph of the linear demand equation is the straight line passing through the points $(1000, 55)$ and $(600, 85)$. The slope of the line is

$$\frac{85 - 55}{600 - 1000} = -\frac{3}{40}.$$

Using the point $(1000, 55)$ and the slope just found, we find that the required equation is $\quad p - 55 = -\frac{3}{40}(x - 1000)$

$$p = -\frac{3}{40}x + 130 \quad .$$

When $x = 0$, $p = 130$ which means that there will be no demand above \$130. When $p = 0$, $x = 1733.33$, which means that 1733 units is the maximum quantity demanded.

39. Since the demand equation is linear, we know that the line passes through the points $(1000, 9)$ and $(6000, 4)$. Therefore, the slope of the line is given by

$$m = \frac{4 - 9}{6000 - 1000} = -\frac{5}{5000} = -0.001.$$

Since the equation of the line has the form $p = ax + b$,

$$9 = -0.001(1000) + b \quad \text{or} \qquad b = 10.$$

Therefore, the equation of the line is

$$p = -0.001x + 10.$$

If $p = 7.50$, $\quad 7.50 = -0.001x + 10$

$$0.001x = 2.50$$

or $\qquad\qquad\qquad x = 2500.$

So, the quantity demanded when the unit price is \$7.50 is 2500 units.

41. a. Setting $x = 0$, we obtain $3(0) - 4p + 24 = 0$

$$-4p = -24$$

or $\qquad\qquad\qquad\qquad p = 6.$

Setting $p = 0$, we obtain $\quad 3x - 4(0) + 24 = 0$

$$3x = -8$$

or $\qquad\qquad\qquad\qquad x = -8/3.$

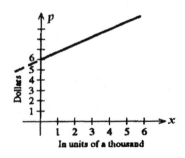

In units of a thousand

b. When $p = 8$,
$$3x - 4(8) + 24 = 0$$
$$3x = 32 - 24 = 8$$
$$x = 8/3.$$

Therefore, 2667 units of the commodity would be supplied at a unit price of $8. (Here again x is measured in units of thousands.)

43. a. When $x = 0$, $p = 10$, and when $p = 0$, $x = -5$.

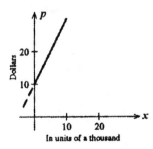

In units of a thousand

b. $p = 2x + 10$, $14 = 2x + 10$, $2x = 4$, and $x = 2$. Therefore, when $p = 14$, the supplier will make 2000 units of the commodity available.

45. When $x = 10,000$, $p = 45$ and when $x = 20,000$, $p = 50$. Therefore, the slope of the line passing $(10,000, 45)$ and $(20,000, 50)$ is

$$m = \frac{50 - 45}{20,000 - 10,000} = \frac{5}{10,000} = 0.0005$$

Using the point- slope form of an equation of a line with the point $(10,000, 45)$, we have
$$p - 45 = 0.0005(x - 10,000)$$
$$p = 0.0005x - 5 + 45$$

1 Straight Lines and Linear Functions

or $\qquad p = 0.0005x + 40.$

If $p = 70,$ $\qquad 70 = 0.0005x + 40$

$\qquad\qquad 0.0005x = 30 \qquad$ or $\qquad\qquad x = \dfrac{30}{0.0005} = 60,000$.

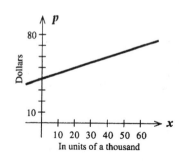

47. False. $P(x) = R(x) - C(x) = sx - (cx + F) = (s - c)x - F.$ Therefore, the firm is

making a profit if $P(x) = (s - c)x - F > 0$ or $x > \dfrac{F}{s - c}.$

USING TECHNOLOGY EXERCISES 1.3, page 45

1. 2.2875 $\qquad$ 3. 2.880952381 $\qquad$ 5. 7.2851648352 $\qquad$ 7. 2.4680851064

EXERCISES 1.4, page 55

1. We solve the system $\quad y = 3x + 4$
$\qquad\qquad\qquad\qquad y = -2x + 14.$
Substituting the first equation into the second yields

$$3x + 4 = -2x + 14$$
$$5x = 10,$$
and $x = 2.$ Substituting this value of x into the first equation yields
$$y = 3(2) + 4,$$
or $y = 10.$ Thus, the point of intersection is $(2, 10).$

3. We solve the system $\quad 2x - 3y = 6$
$\qquad\qquad\qquad\qquad 3x + 6y = 16.$
Solving the first equation for $y,$ we obtain
$$3y = 2x - 6$$
$$y = \tfrac{2}{3}x - 2 \ .$$

Substituting this value of y into the second equation, we obtain
$$3x + 6(\tfrac{2}{3}x - 2) = 16$$
$$3x + 4x - 12 = 16$$
$$7x = 28$$
and $\qquad\qquad x = 4.$

Then $\qquad\quad y = \tfrac{2}{3}(4) - 2 = \tfrac{2}{3}.$

Therefore, the point of intersection is $(4, \tfrac{2}{3})$.

5. We solve the system $\begin{cases} y = \tfrac{1}{4}x - 5 \\ 2x - \tfrac{3}{2}y = 1 \end{cases}$. Substituting the value of y given in the first

equation into the second equation, we obtain
$$2x - \tfrac{3}{2}(\tfrac{1}{4}x - 5) = 1$$
$$2x - \tfrac{3}{8}x + \tfrac{15}{2} = 1$$
$$16x - 3x + 60 = 8$$
$$13x = -52,$$
or $x = -4$. Substituting this value of x in the first equation, we have
$$y = \tfrac{1}{4}(-4) - 5 = -1 - 5,$$
or $y = -6$. Therefore, the point of intersection is $(-4, -6)$.

7. We solve the equation $R(x) = C(x)$, or $15x = 5x + 10{,}000$, obtaining $10x = 10{,}000$, or $x = 1000$. Substituting this value of x into the equation $R(x) = 15x$, we find $R(1000) = 15{,}000$. Therefore, the breakeven point is $(1000, 15{,}000)$.

9. We solve the equation $R(x) = C(x)$, or $0.4x = 0.2x + 120$, obtaining $0.2x = 120$, or $x = 600$. Substituting this value of x into the equation $R(x) = 0.4x$, we find $R(600) = 240$. Therefore, the breakeven point is $(600, 240)$.

11. a.

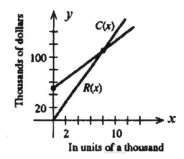

b. We solve the equation $R(x) = C(x)$ or $14x = 8x + 48{,}000$, obtaining $6x = 48{,}000$ or $x = 8000$. Substituting this value of x into the equation $R(x) = 14x$, we find $R(8000) = 14(8000) = 112{,}000$. Therefore, the breakeven point is $(8000, 112{,}000)$.

c.

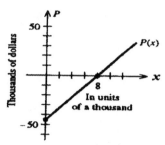

d. $P(x) = R(x) - C(x) = 14x - 8x - 48{,}000 = 6x - 48{,}000$.
The graph of the profit function crosses the x-axis when $P(x) = 0$, or $6x = 48{,}000$ and $x = 8000$. This means that the revenue is equal to the cost when 8000 units are produced and consequently the company breaks even at this point.

13. Let x denote the number of units sold. Then, the revenue function R is given by
$$R(x) = 9x.$$
Since the variable cost is 40 percent of the selling price and the monthly fixed costs are $50,000, the cost function C is given by
$$C(x) = 0.4(9x) + 50{,}000$$
$$= 3.6x + 50{,}000.$$
To find the breakeven point, we set $R(x) = C(x)$, obtaining
$$9x = 3.6x + 50{,}000$$
$$5.4x = 50{,}000$$
$$x \approx 9259\text{, or 9259 units.}$$
Substituting this value of x into the equation $R(x) = 9x$ gives
$$R(9259) = 9(9259) = 83{,}331.$$

Thus, for a breakeven operation, the firm should manufacture 9259 bicycle pumps resulting in a breakeven revenue of $83,331.

15. a. The cost function associated with using machine I is given by
$$C_1(x) = 18{,}000 + 15x.$$
The cost function associated with using machine II is given by
$$C_2(x) = 15{,}000 + 20x.$$

b.

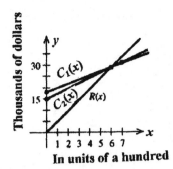

In units of a hundred

c. Comparing the cost of producing 450 units on each machine, we find
$$C_1(450) = 18{,}000 + 15(450)$$
$$= 24{,}750 \quad \text{or } \$24{,}750 \text{ on machine } I,$$
and
$$C_2(450) = 15{,}000 + 20(450)$$
$$= 24{,}000 \text{ or } \$24{,}000 \text{ on machine } II.$$
Therefore, machine II should be used in this case.
Next, comparing the costs of producing 550 units on each machine, we find
$$C_1(550) = 18{,}000 + 15(550)$$
$$= 26{,}250 \text{ or } \$26{,}250 \text{ on machine } I,$$
and
$$C_2(550) = 15{,}000 + 20(550)$$
$$= 26{,}000$$
on machine II. Therefore, machine II should be used in this instance. Once again, we compare the cost of producing 650 units on each machine and find that
$$C_1(650) = 18{,}000 + 15(650)$$
$$= 27{,}750, \quad \text{or } \$27{,}750 \text{ on machine } I \text{ and}$$
$$C_2(650) = 15{,}000 + 20(650)$$
$$= 28{,}000,$$
or \$28,000 on machine II. Therefore, machine I should be used in this case.

d. We use the equation $P(x) = R(x) - C(x)$ and find
$$P(450) = 50(450) - 24{,}000 = -1500,$$
or a loss of \$1500 when machine I is used to produce 450 units. Similarly,
$$P(550) = 50(550) - 26{,}000 = 1500,$$
or a profit of \$1500 when machine II is used to produce 550 units.
Finally, $P(650) = 50(650) - 27{,}750 = 4750,$
or a profit of \$4750 when machine I is used to produce 650 units.

17. We solve the system
$$4x + 3p = 59$$
$$5x - 6p = -14.$$

21

Solving the first equation for p, we find $p = -\frac{4}{3}x + \frac{59}{3}$.

Substituting this value of p into the second equation, we have
$$5x - 6(-\tfrac{4}{3}x + \tfrac{59}{3}) = -14$$
$$5x + 8x - 118 = -14$$
$$13x = 104$$
$$x = 8.$$

Substituting this value of x into the equation
$$p = -\tfrac{4}{3}x + \tfrac{59}{3}$$

we have
$$p = -\tfrac{4}{3}(8) + \tfrac{59}{3} = \tfrac{27}{3} = 9$$

Thus, the equilibrium quantity is 8000 units and the equilibrium price is $9.

19. We solve the system $p = -2x + 22$
$$p = 3x + 12 .$$

Substituting the first equation into the second, we find
$$-2x + 22 = 3x + 12$$
$$5x = 10$$

and $x = 2.$

Substituting this value of x into the first equation, we obtain
$$p = -2(2) + 22 = 18.$$

Thus, the equilibrium quantity is 2000 units and the equilibrium price is $18.

21. Let x denote the number of VCR's produced per week, and p denote the price of each VCR.

a. The slope of the demand curve is given by $\dfrac{\Delta p}{\Delta x} = -\dfrac{20}{250} = -\dfrac{2}{25}$.

Using the point-slope form of the equation of a line with the point $(3000, 485)$, we have $\quad p - 485 = -\tfrac{2}{25}(x - 3000)$
$$p = -\tfrac{2}{25}x + 240 + 485$$

or $\quad\quad p = -0.08x + 725.$

b. From the given information, we know that the graph of the supply equation passes through the points $(0, 300)$ and $(2500, 525)$. Therefore, the slope of the supply curve

is $\quad\quad m = \dfrac{525 - 300}{2500 - 0} = \dfrac{225}{2500} = 0.09$.

Using the point-slope form of the equation of a line with the point $(0, 300)$, we find that $\quad p - 300 = 0.09x$
$$p = 0.09x + 300.$$

c. Equating the supply and demand equations, we have
$$-0.08x + 725 = 0.09x + 300$$
$$0.17x = 425$$
or
$$x = 2500.$$
Then
$$p = -0.08(2500) + 725 = 525.$$
We conclude that the equilibrium quantity is 2500 and the equilibrium price is $525.

23. We solve the system
$$3x + p = 1500$$
$$2x - 3p = -1200.$$
Solving the first equation for p, we obtain
$$p = 1500 - 3x.$$
Substituting this value of p into the second equation, we obtain
$$2x - 3(1500 - 3x) = -1200$$
$$11x = 3300$$
or
$$x = 300.$$
Next,
$$p = 1500 - 3(300) = 600.$$
Thus, the equilibrium quantity is 300 and the equilibrium price is $600.

25. a. We solve the system of equations $p = cx + d$ and $p = ax + b$. Substituting the first into the second gives
$$cx + d = ax + b$$
$$(c - a)x = b - d$$
or
$$x = \frac{b-d}{c-a}.$$
Since $a < 0$ and $c > 0$, $c - a \neq 0$ and x is well-defined. Substituting this value of x into the second equation, we obtain
$$p = a\left(\frac{b-d}{c-a}\right) + b = \frac{ab - ad + bc - ab}{c-a} = \frac{bc - ad}{c-a}.$$

Therefore, the equilibrium quantity is $\frac{b-d}{c-a}$ and the equilibrium price is $\frac{bc-ad}{c-a}$.

b. If c is increased, the denominator in the expression for x increases and so x gets smaller. At the same time, the first term in the first equation for p decreases and so p gets larger. This analysis shows that if the unit price for producing the product is increased then the equilibrium quantity decreases while the equilibrium price increases.

c. If b is decreased, the numerator of the expression for x decreases while the denominator stays the same. Therefore x decreases. The expression for p also shows that p decreases. This analysis shows that if the (theoretical) upper bound for the unit price of a commodity is lowered, then both the equilibrium quantity and the equilibrium price drop.

27. True. $P(x) = R(x) - C(x) = sx - (cx + F) = (s - c)x - F$. Therefore, the firm is making a profit if $P(x) = (s - c)x - F > 0$, or $x > \frac{F}{s-c}$ $(s \neq c)$.

29. Solving the two equations simultaneously to find the point(s) of intersection of L_1 and L_2, we obtain
$$m_1x + b_1 = m_2x + b_2$$
$$(m_1 - m_2)x = b_2 - b_1 \tag{1}$$
a. If $m_1 = m_2$ and $b_2 \neq b_1$, then there is no solution for (1) and in this case L_1 and L_2 do not intersect.
b. If $m_1 \neq m_2$, then Equation (1) can be solved (uniquely) for x and this shows that L_1 and L_2 intersect at precisely one point.
c. If $m_1 = m_2$ and $b_1 = b_2$, then (1) is satisfied for all values of x and this shows that L_1 and L_2 intersect at infinitely many points.

USING TECHNOLOGY EXERCISES 1.4, page 59

1. (0.6, 6.2)

3. (3.8261, 0.1304)

5. (386.9091, 145.3939)

EXERCISES 1.5, page 65

1. a. We first summarize the data:

x	y	x^2	xy
1	4	1	4
2	6	4	12
3	8	9	24
4	11	16	44
10	29	30	84

The normal equations are
$$4b + 10m = 29$$
$$10b + 30m = 84.$$
Solving this system of equations, we obtain $m = 2.3$ and $b = 1.5$. So an equation is $y = 2.3x + 1.5$.

b. The scatter diagram and the least squares line for this data follow:

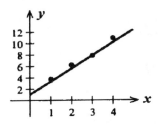

3. a. We first summarize the data:

x	y	x^2	xy
1	4.5	1	4.5
2	5	4	10
3	3	9	9
4	2	16	8
4	3.5	16	14
6	1	36	6
20	19	82	51.5

The normal equations are $6b + 20m = 19$
$$20b + 82m = 51.5.$$
The solutions are $m \approx -0.7717$ and $b \approx 5.7391$ and so a required equation is
$y = -0.772x + 5.739$.
b. The scatter diagram and the least-squares line for these data follow.

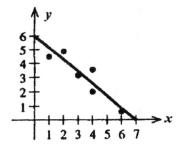

5. a. We first summarize the data:

x	y	x^2	xy
1	3	1	3
2	5	4	10
3	5	9	15
4	7	16	28
5	8	25	40
15	28	55	96

The normal equations are $55m + 15b = 96$
$$15m + 5b = 28.$$
Solving, we find $m = 1.2$ and $b = 2$, so that the required equation is $y = 1.2x + 2$.
b.   The scatter diagram and the least-squares line for the given data follow.

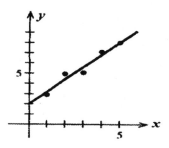

7. a. We first summarize the data:

x	y	x^2	xy
4	0.5	16	2
4.5	0.6	20.25	2.7
5	0.8	25	4
5.5	0.9	30.25	4.95
6	1.2	36	7.2
25	4	127.5	20.85

The normal equations are $5b + 25m = 4$
$$25b + 127.5m = 20.85.$$
The solutions are $m = 0.34$ and $b = -0.9$, and so a required equation is
$y = 0.34x - 0.9$.
b. The scatter diagram and the least-squares line for these data follow.

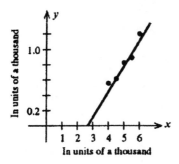

In units of a thousand

In units of a thousand

c. If $x = 6.4$, then $y = 0.34(6.4) - 0.9 = 1.276$ and so 1276 completed applications might be expected.

9.   a. We first summarize the data:

x	y	x^2	xy
1	436	1	436
2	438	4	876
3	428	9	1284
4	430	16	1720
5	426	25	2130
15	2158	55	6446

The normal equations are
$$5b + 15m = 2158$$
$$15b + 55m = 6446.$$
Solving this system, we find $m = -2.8$ and $b = 440$.
Thus, the equation of the least-squares line is $y = -2.8x + 440$.
b. The scatter diagram and the least-squares line for this data are shown in the figure that follows.

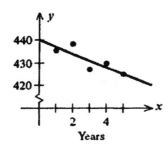

Years

1 Straight Lines and Linear Functions

c. Two years from now, the average SAT verbal score in that area will be $y = -2.8(7) + 440 = 420.4$.

11. a. We first summarize the data:

x	y	x^2	xy
0	168	0	0
10	213	100	2130
20	297	400	5940
30	374	900	11220
40	427	1600	17080
57	471	3249	26847
157	1950	6249	63217

The normal equations are $\quad 6b + 157m = \quad 1950$

$$157b + 6249m = 63217.$$

The solutions are $m = 5.695 \approx 5.70$ and $b = 175.98 \approx 176$ and so a required equation is $y = 5.70x + 176$.

b. In 2003, $x = 63$, $y = 5.70\,(63) + 176 = 535.10 \approx 535$. Hence, the expected size of the average farm will be 535 acres.

13. a. We first summarize the data:

x	y	x^2	xy
1	20	1	20
2	24	4	48
3	26	9	78
4	28	16	112
5	32	25	160
15	130	55	418

The normal equations are $\quad 5b + 15m = 130$
$$15b + 55m = 418.$$

The solutions are $m = 2.8$ and $b = 17.6$, and so an equation of the line is
$$y = 2.8x + 17.6.$$

b. When $x = 8$, $y = 2.8(8) + 17.6 = 40$. Hence, the state subsidy is expected to be $40 million for the eighth year.

15. a. We first summarize the data:

x	y	x^2	xy
1	16.7	1	16.7
3	26	9	78
5	33.3	25	166.5
7	48.3	49	338.1
9	57	81	513
11	65.8	121	723.8
13	74.2	169	964.6
15	83.3	225	1249.5
64	404.6	680	4050.2

The normal equations are $8b + 64m = 404.6$

$$64b + 680m = 4050.2.$$

The solutions are $m = 4.8417$ and $b = 11.8417$ and so a required equation is $y = 4.842x + 11.842$.

b. In 1993, $x = 19$, and so $y = 4.842(19) + 11.842 = 103.84$. Hence the estimated number of cans produced in 1993 is 103.8 billion.

17. a. We first summarize the data:

x	y	x^2	xy
0	21.7	0	0
1	32.1	1	32.1
2	45.0	4	90
3	58.3	9	174.9
4	69.6	16	278.4
10	226.7	30	575.4

The normal equations are $\begin{aligned} 5b + 10m &= 226.7 \\ 10b + 30m &= 575.4 \end{aligned}$.

The solutions are $m = 12.2$ and $b = 20.9$ and so a required equation is $y = 12.2x + 20$.

b. In 2003, $x = 5$ so $y = 12.2(5) + 20.9 = 81.9$, or 81.9 million computers are expected to be connected to the internet in Europe in that year.

19. a. We first summarize the data:

x	y	x^2	xy
4.25	178	18.0625	756.5
10	667	100	6670
14	1194	196	16716
15.5	1500	240.25	23250
17.8	1388	316.84	24706.4
19.5	1640	380.25	31980
81.05	6567	1251.4025	104,078.9

The normal equations are $\quad 6b + \qquad 81.05m = 6567$
$$81.05b + \quad 1251.4025m = 104{,}078.9.$$
The solutions are $m = 98.1761$ and $b = -231.696$ and so a required equation is
$y = 98.176x - 231.7$.
b. If $x = 20$, then $y = 98.176(20) - 231.7 = 1731.82$. Hence, if the health-spending in the U.S. were in line with OECD countries, it should only have been $1732 per capita.

21. True, The error involves the sum of the squares of the form $[f(x_i) - y_i]^2$ where f is the least-squares function and y_i is a data point. Thus, the error is zero if and only if $f(x_i) = y_i$ for each $1 \le i \le n$.

USING TECHNOLOGY EXERCISES 1.5, page 69

1. $y = 2.3596x + 3.8639$ 3. $y = -1.1948x + 3.5525$

5. a. $y = 13.321x + 72.57$ b. 192 million tons

CHAPTER 1 REVIEW EXERCISES, page 73

1. The distance is $\quad d = \sqrt{(6-2)^2 + (4-1)^2} = \sqrt{4^2 + 3^2} = \sqrt{25} = 5$.

3. The distance is
$$d = \sqrt{[1-(-2)]^2 + [-7-(-3)]^2} = \sqrt{3^2 + (-4)^2} = \sqrt{9+16} = \sqrt{25} = 5.$$

5. An equation is $x = -2$.

7. The slope of L is $m = \dfrac{\frac{7}{2}-4}{3-(-2)} = -\dfrac{1}{10}$ and an equation of L is
$$y-4 = -\tfrac{1}{10}\left[x-(-2)\right] = -\tfrac{1}{10}x - \tfrac{1}{5},$$
or $\qquad y = -\tfrac{1}{10}x + \tfrac{19}{5}.$

The general form of this equation is $x + 10y - 38 = 0$.

9. Writing the given equation in the form $y = \tfrac{5}{2}x - 3$, we see that the slope of the given line is 5/2. So a required equation is
$$y-4 = \tfrac{5}{2}(x+2) \quad \text{or} \quad y = \tfrac{5}{2}x + 9.$$

The general form of this equation is $5x - 2y + 18 = 0$.

11. Using the slope-intercept form of the equation of a line, we have $y = -\tfrac{1}{2}x - 3$.

13. Rewriting the given equation in the slope-intercept form, we have $4y = -3x + 8$
or $\qquad y = -\tfrac{3}{4}x + 2$
and conclude that the slope of the required line is $-3/4$. Using the point-slope form of the equation of a line with the point $(2,3)$ and slope $-3/4$, we obtain
$$y - 3 = -\tfrac{3}{4}(x-2)$$
$$y = -\tfrac{3}{4}x + \tfrac{6}{4} + 3$$
$$= -\tfrac{3}{4}x + \tfrac{9}{2}.$$
The general form of this equation is $3x + 4y - 18 = 0$.

15. Rewriting the given equation in the slope-intercept form $y = \tfrac{2}{3}x - 8$, we see that the slope of the line with this equation is 2/3. The slope of the required line is $-3/2$. Using the point-slope form of the equation of a line with the point $(-2,-4)$ and slope $-3/2$, we have
$$y - (-4) = -\tfrac{3}{2}\left[x - (-2)\right]$$
or $\qquad y = -\tfrac{3}{2}x - 7$.
The general form of this equation is $3x + 2y + 14 = 0$.

17. Setting $x = 0$, gives $5y = 15$, or $y = 3$. Setting $y = 0$, gives $-2x = 15$, or $x = -15/2$. The graph of the equation $-2x + 5y = 15$ follows.

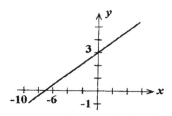

19. Let x denote the time in years. Since the function is linear, we know that it has the form $f(x) = mx + b$.

    a. The slope of the line passing through (0, 2.4) and (5, 7.4) is

$$m = \frac{7.4 - 2.4}{5} = 1.$$

Since the line passes through (0, 2.4), we know that the y-intercept is 2.4. Therefore, the required function is $f(x) = x + 2.4$.

    b. In 1997 ($x = 3$), the sales were $f(3) = 3 + 2.4 = 5.4$, or \$5.4 million dollars.

21. a. $D(w) = \dfrac{a}{150} w$. The given equation can be expressed in the form $y = mx + b$,

    where $m = \dfrac{a}{150}$ and $b = 0$.

    b. If $a = 500$ and $w = 35$, $D(35) = \frac{500}{150}(35) = 116\frac{2}{3}$, or approximately 117 mg.

23. Let V denote the value of the machine after t years.

    a. The rate of depreciation is

$$-\frac{\Delta V}{\Delta t} = \frac{300,000 - 30,000}{12} = \frac{270,000}{12} = 22,500, \quad \text{or } \$22,500/\text{year}.$$

    b. Using the point-slope form of the equation of a line with the point (0, 300,000) and $m = -22,500$, we have

$$V - 300,000 = -22,500(t - 0)$$
$$V = -22,500t + 300,000.$$

25. The slope of the demand curve is $\dfrac{\Delta p}{\Delta x} = -\dfrac{10}{200} = -0.05$.

    Using the point-slope form of the equation of a line with the point (0, 200), we have

$$p - 200 = -0.05(x), \quad \text{or} \quad p = -0.05x + 200.$$

The graph of the demand equation follows.

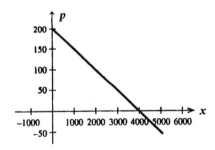

27. We solve the system $3x + 4y = -6$
$$2x + 5y = -11.$$
Solving the first equation for x, we have $3x = -4y - 6$
and $$x = -\tfrac{4}{3}y - 2.$$
Substituting this value of x into the second equation yields
$$2(-\tfrac{4}{3}y - 2) + 5y = -11$$
$$-\tfrac{8}{3}y - 4 + 5y = -11$$
$$\tfrac{7}{3}y = -7, \quad \text{or} \quad y = -3.$$
Then $$x = -\tfrac{4}{3}(-3) - 2 = 4 - 2 = 2.$$
Therefore, the point of intersection is $(2, -3)$.

29. Setting $C(x) = R(x)$, we have $12x + 20{,}000 = 20x$
$$8x = 20{,}000$$
or $$x = 2500.$$
Next, $$R(2500) = 20(2500) = 50{,}000,$$
and we conclude that the breakeven point is $(2500, 50{,}000)$.

31. a. The slope of the line is $m = \dfrac{1 - 0.5}{4 - 2} = 0.25$.

Using the point-slope form of an equation of a line, we have
$$y - 1 = 0.25(x - 4)$$
$$y = 0.25x$$
b. $$y = 0.25(6.4) = 1.6, \quad \text{or } 1600 \text{ applications.}$$

CHAPTER 2

EXERCISES 2.1, page 84

1. Solving the first equation for x, we find $x = 3y - 1$. Substituting this value of x into the second equation yields
$$4(3y - 1) + 3y = 11$$
$$12y - 4 + 3y = 11$$
or
$$y = 1.$$
Substituting this value of y into the first equation gives
$$x = 3(1) - 1 = 2$$
Therefore, the unique solution of the system is $(2,1)$.

3. Solving the first equation for x, we have $x = 7 - 4y$. Substituting this value of x into the second equation, we have
$$\tfrac{1}{2}(7 - 4y) + 2y = 5$$
$$7 - 4y + 4y = 10$$
$$7 = 10.$$
Clearly, this is impossible and we conclude that the system of equations has no solution.

5. Solving the first equation for x, we obtain $x = 7 - 2y$.
Substituting this value of x into the second equation, we have
$$2(7 - 2y) - y = 4$$
$$14 - 4y - y = 4$$
$$-5y = -10$$
and
$$y = 2.$$
Then
$$x = 7 - 2(2) = 7 - 4 = 3.$$
We conclude that the solution to the system is $(3,2)$.

7. Solving the first equation for x, we have
$$2x = 5y + 10$$
and
$$x = \tfrac{5}{2}y + 5.$$
Substituting this value of x into the second equation, we have
$$6(\tfrac{5}{2}y + 5) - 15y = 30$$
$$15y + 30 - 15y = 30$$
or
$$0 = 0.$$

This result tells us that the second equation is equivalent to the first. Thus, any ordered pair of numbers (x, y) satisfying the equation

$$2x - 5y = 10 \qquad \text{(or } 6x - 15y = 30)$$

is a solution to the system. In particular, by assigning the value t to x, where t is any real number, we find that

$$y = -2 + \tfrac{2}{5}t$$

so the ordered pair, $(t, \tfrac{2}{5}t - 2)$ is a solution to the system, and we conclude that the system has infinitely many solutions.

9. Solving the first equation for x, we obtain

$$4x - 5y = 14$$
$$4x = 14 + 5y$$
$$x = \tfrac{14}{4} + \tfrac{5}{4}y = \tfrac{7}{2} + \tfrac{5}{4}y.$$

Substituting this value of x into the second equation gives

$$2(\tfrac{7}{2} + \tfrac{5}{4}y) + 3y = -4$$
$$7 + \tfrac{5}{2}y + 3y = -4$$
$$\tfrac{11}{2}y = -11$$

or
$$y = -2.$$
Then,
$$x = \tfrac{7}{2} + \tfrac{5}{4}(-2) = 1.$$

We conclude that the ordered pair $(1, -2)$ satisfies the given system of equations.

11. Solving the first equation for x, we obtain

$$2x = 3y + 6$$
$$x = \tfrac{3}{2}y + 3$$

Substituting this value of x into the second equation gives

$$6(\tfrac{3}{2}y + 3) - 9y = 12$$
$$9y + 18 - 9y = 12$$
$$18 = 12.$$

which is impossible. We conclude that the system of equations has no solution.

13. Solving the first equation for y, we obtain $y = 2x - 3$. Substituting this value of y into the second equation yields

$$4x + k(2x - 3) = 4,$$
$$4x + 2xk - 3k = 4$$
$$2x(2 + k) = 4 + 3k$$
$$x = \frac{4 + 3k}{2(2 + k)}.$$

Since x is not defined when the denominator of this last expression is zero, we conclude that the system has no solution when $k = -2$.

15. Let x and y denote the number of acres of corn and wheat planted, respectively. Then $x + y = 500$. Since the cost of cultivating corn is \$42/acre and that of wheat \$30/acre and Mr. Johnson has \$18,600 available for cultivation, we have $42x + 30y = 18600$. Thus, the solution is found by solving the system of equations

$$\begin{aligned} x + \quad y &= \quad 500 \\ 42x + 30y &= 18{,}600 \end{aligned}$$

17. Let x denote the number of pounds of the \$2.50/lb coffee and y denote the number of pounds of the \$3/lb coffee. Then

$$x + y = 100.$$

Since the blended coffee sells for \$2.80/lb, we know that the blended mixture is worth $(2.80)(100) = \$280$. Therefore,

$$2.50x + 3y = 280.$$

Thus, the solution is found by solving the system of equations

$$\begin{aligned} x + \quad y &= 100 \\ 2.50x + 3y &= 280 \end{aligned}$$

19. Let x denote the number of children who rode the bus during the morning shift and y denote the number of adults who rode the bus during the morning shift. Then $x + y = 1000$. Since the total fare collected was \$650, we have $0.25x + 0.75y = 650$. Thus, the solution to the problem can be found by solving the system of equations

$$\begin{aligned} x + \quad y &= 1000 \\ 0.25x + 0.75y &= \quad 650. \end{aligned}$$

21. Let $x =$ the amount of money invested at 6 percent in a savings account

$\quad\quad\quad y =$ the amount of money invested at 8 percent in mutual funds

and $\quad z =$ the amount of money invested at 12 percent in bonds.

Since the total interest was \$21,600, we have

$$0.6x + 0.8y + 0.12z = 21{,}600.$$

Also, since the amount of Mr. Carrington's investment in bonds is twice the amount of the investment in the savings account, we have

$$z = 2x.$$

Finally, the interest earned from his investment in bonds was equal to the interest earned on his money market certificates, so

$$0.08y = 0.12z.$$

Thus, the solution to the problem can be found by solving the system of equations

$$0.06x + 0.08y + 0.12z = 21,600$$
$$2x \qquad - \quad z = \qquad 0$$
$$0.08y - 0.12z = \qquad 0.$$

23. Let x, y, and z denote the number of compact, intermediate, and full-size cars, respectively, to be purchased. The cost incurred in buying the specified number of cars is $10000x + 15000y + 20000z$. Since the budget is $1.25 million, we have the system

$$10,000x + 15,000y + 20,000z = 1,250,000$$
$$x - \qquad 2y \qquad = \qquad 0$$
$$x + \qquad y + \qquad z = \qquad 100.$$

25. Let $\quad x =$ the number of ounces of Food I used in the meal
$\qquad y =$ the number of ounces of Food II used in the meal
and $\quad z =$ the number of ounces of Food III used in the meal.
Since 100 percent of the daily requirement of proteins, carbohydrates, and iron is to be met by this meal, we have the following system of linear equations:

$$10x + 6y + 8z = 100$$
$$10x + 12y + 6z = 100$$
$$5x + 4y + 12z = 100.$$

27. True. If the three lines coincide, then the system has infinitely many solutions-- corresponding to all the points on the (common) line. If at least one line is distinct from the others, then the system has no solution.

EXERCISES 2.2, page 98

1. $\begin{bmatrix} 2 & -3 & | & 7 \\ 3 & 1 & | & 4 \end{bmatrix}$

3. $\begin{bmatrix} 0 & -1 & 2 & | & 6 \\ 2 & 2 & -8 & | & 7 \\ 0 & 3 & 4 & | & 0 \end{bmatrix}$

5. $3x + 2y = -4$
$\quad x - y = 5$

7. $\quad x + 3y + 2z = 4$
$\quad 2x \qquad\qquad = 5$
$\quad 3x - 3y + 2z = 6$

9. Yes. Conditions 1-4 are satisfied (see page 91 of the text).

11. No. Condition 3 is violated. The first nonzero entry in the second row does not lie to the right of the first nonzero entry 1 in row 1.

13. Yes. Conditions 1-4 are satisfied.

15. No. Condition 2 and consequently condition 4 are not satisfied. The first nonzero entry in the last row is not a 1 and the column containing that entry does not have zeros elsewhere.

17. No. Condition 1 is violated. The first row consists entirely of zeros and it lies above row 2.

19. $\begin{bmatrix} \boxed{2} & 4 & | & 8 \\ 3 & 1 & | & 2 \end{bmatrix} \xrightarrow{\frac{1}{2}R_1} \begin{bmatrix} 1 & 2 & | & 4 \\ 3 & 1 & | & 2 \end{bmatrix} \xrightarrow{R_2-3R_1} \begin{bmatrix} 1 & 2 & | & 4 \\ 0 & -5 & | & -10 \end{bmatrix}$

21. $\begin{bmatrix} \boxed{-1} & 2 & | & 3 \\ 6 & 4 & | & 2 \end{bmatrix} \xrightarrow{-R_1} \begin{bmatrix} 1 & -2 & | & -3 \\ 6 & 4 & | & 2 \end{bmatrix} \xrightarrow{R_2-6R_1} \begin{bmatrix} 1 & -2 & | & -3 \\ 0 & 16 & | & 20 \end{bmatrix}$

23. $\begin{bmatrix} \boxed{2} & 4 & 6 & | & 12 \\ 2 & 3 & 1 & | & 5 \\ 3 & -1 & 2 & | & 4 \end{bmatrix} \xrightarrow{\frac{1}{2}R_1} \begin{bmatrix} 1 & 2 & 3 & | & 6 \\ 2 & 3 & 1 & | & 5 \\ 3 & -1 & 2 & | & 4 \end{bmatrix} \xrightarrow[R_3-3R_1]{R_2-2R_1} \begin{bmatrix} 1 & 2 & 3 & | & 6 \\ 0 & -1 & -5 & | & -7 \\ 0 & -7 & -7 & | & -14 \end{bmatrix}$

25. $\begin{bmatrix} 0 & 1 & 3 & | & 4 \\ 2 & 4 & \boxed{1} & | & 3 \\ 5 & 6 & 2 & | & -4 \end{bmatrix} \xrightarrow[R_3-2R_2]{R_1-3R_2} \begin{bmatrix} -6 & -11 & 0 & | & -5 \\ 2 & 4 & 1 & | & 3 \\ 1 & -2 & 0 & | & -10 \end{bmatrix}$

27. $\begin{bmatrix} \boxed{3} & 9 & | & 6 \\ 2 & 1 & | & 4 \end{bmatrix} \xrightarrow{\frac{1}{3}R_1} \begin{bmatrix} 1 & 3 & | & 2 \\ 2 & 1 & | & 4 \end{bmatrix} \xrightarrow{R_2-2R_1} \begin{bmatrix} 1 & 3 & | & 2 \\ 0 & -5 & | & 0 \end{bmatrix} \xrightarrow{-\frac{1}{5}R_2}$

$\begin{bmatrix} 1 & 3 & | & 2 \\ 0 & 1 & | & 0 \end{bmatrix} \xrightarrow{R_1-3R_2} \begin{bmatrix} 1 & 0 & | & 2 \\ 0 & 1 & | & 0 \end{bmatrix}$

29. $\begin{bmatrix} \boxed{1} & 3 & 1 & | & 3 \\ 3 & 8 & 3 & | & 7 \\ 2 & -3 & 1 & | & -10 \end{bmatrix} \xrightarrow[R_3-2R_1]{R_2-3R_1} \begin{bmatrix} 1 & 3 & 1 & | & 3 \\ 0 & -1 & 0 & | & -2 \\ 0 & -9 & -1 & | & -16 \end{bmatrix} \xrightarrow{-R_2} \begin{bmatrix} 1 & 3 & 1 & | & 3 \\ 0 & 1 & 0 & | & 2 \\ 0 & -9 & -1 & | & -16 \end{bmatrix}$

$$\xrightarrow[\substack{R_1-3R_2 \\ R_3+9R_2}]{} \begin{bmatrix} 1 & 0 & 1 & -3 \\ 0 & 1 & 0 & 2 \\ 0 & 0 & -1 & 2 \end{bmatrix} \xrightarrow[\substack{R_1+R_3 \\ -R_3}]{} \begin{bmatrix} 1 & 0 & 0 & -1 \\ 0 & 1 & 0 & 2 \\ 0 & 0 & 1 & -2 \end{bmatrix}.$$

31. The augmented matrix is equivalent to the system of linear equations

$$3x + 9y = 6$$
$$2x + y = 4$$

The ordered pair $(2,0)$ is the solution to the system.

33. The augmented matrix is equivalent to the system of linear equations

$$x + 3y + z = 3$$
$$3x + 8y + 3z = 7$$
$$2x - 3y + z = -10$$

Reading off the solution from the last augmented matrix,

$$\begin{bmatrix} 1 & 0 & 0 & -1 \\ 0 & 1 & 0 & 2 \\ 0 & 0 & 1 & -2 \end{bmatrix},$$

which is in row-reduced form, we have $x = -1$, $y = 2$, and $z = -2$.

35. Using the Gauss-Jordan method, we have

$$\begin{bmatrix} 1 & -2 & 8 \\ 3 & 4 & 4 \end{bmatrix} \xrightarrow{R_2-3R_1} \begin{bmatrix} 1 & -2 & 8 \\ 0 & 10 & -20 \end{bmatrix} \xrightarrow{\frac{1}{10}R_2} \begin{bmatrix} 1 & -2 & 8 \\ 0 & 1 & -2 \end{bmatrix} \xrightarrow{R_1+2R_2} \begin{bmatrix} 1 & 0 & 4 \\ 0 & 1 & -2 \end{bmatrix}$$

The solution is $(4,-2)$.

37. Using the Gauss-Jordan method, we have

$$\begin{bmatrix} 2 & -3 & -8 \\ 4 & 1 & -2 \end{bmatrix} \xrightarrow{\frac{1}{2}R_1} \begin{bmatrix} 1 & -\frac{3}{2} & -4 \\ 4 & 1 & -2 \end{bmatrix} \xrightarrow{R_2-4R_1} \begin{bmatrix} 1 & -\frac{3}{2} & -4 \\ 0 & 7 & 14 \end{bmatrix} \xrightarrow{\frac{1}{7}R_2}$$

$$\begin{bmatrix} 1 & -\frac{3}{2} & -4 \\ 0 & 1 & 2 \end{bmatrix} \xrightarrow{R_1+\frac{3}{2}R_2} \begin{bmatrix} 1 & 0 & -1 \\ 0 & 1 & 2 \end{bmatrix}.$$

The solution is $(-1,2)$.

39. Using the Gauss-Jordan method, we have

$$\begin{bmatrix} 1 & 1 & 1 & | & 0 \\ 2 & -1 & 1 & | & 1 \\ 1 & 1 & -2 & | & 2 \end{bmatrix} \xrightarrow[R_3-R_1]{R_2-2R_1} \begin{bmatrix} 1 & 1 & 1 & | & 0 \\ 0 & -3 & -1 & | & 1 \\ 0 & 0 & -3 & | & 2 \end{bmatrix} \xrightarrow{-\frac{1}{3}R_2} \begin{bmatrix} 1 & 1 & 1 & | & 0 \\ 0 & 1 & \frac{1}{3} & | & -\frac{1}{3} \\ 0 & 0 & -3 & | & 2 \end{bmatrix} \xrightarrow{R_1-R_2}$$

$$\begin{bmatrix} 1 & 0 & \frac{2}{3} & | & \frac{1}{3} \\ 0 & 1 & \frac{1}{3} & | & -\frac{1}{3} \\ 0 & 0 & -3 & | & 2 \end{bmatrix} \xrightarrow{-\frac{1}{3}R_3} \begin{bmatrix} 1 & 0 & \frac{2}{3} & | & \frac{1}{3} \\ 0 & 1 & \frac{1}{3} & | & -\frac{1}{3} \\ 0 & 0 & 1 & | & -\frac{2}{3} \end{bmatrix} \xrightarrow[R_2-\frac{1}{3}R_3]{R_1-\frac{2}{3}R_3} \begin{bmatrix} 1 & 0 & 0 & | & \frac{7}{9} \\ 0 & 1 & 0 & | & -\frac{1}{9} \\ 0 & 0 & 1 & | & -\frac{2}{3} \end{bmatrix}.$$

The solution is $\left(\frac{7}{9},-\frac{1}{9},-\frac{2}{3}\right)$.

41. $\begin{bmatrix} 2 & 2 & 1 & | & 9 \\ 1 & 0 & 1 & | & 4 \\ 0 & 4 & -3 & | & 17 \end{bmatrix} \xrightarrow{R_1 \leftrightarrow R_2} \begin{bmatrix} 1 & 0 & 1 & | & 4 \\ 2 & 2 & 1 & | & 9 \\ 0 & 4 & -3 & | & 17 \end{bmatrix} \xrightarrow{R_2-2R_1} \begin{bmatrix} 1 & 0 & 1 & | & 4 \\ 0 & 2 & -1 & | & 1 \\ 0 & 4 & -3 & | & 17 \end{bmatrix} \xrightarrow{\frac{1}{2}R_2}$

$\begin{bmatrix} 1 & 0 & 1 & | & 4 \\ 0 & 1 & -\frac{1}{2} & | & \frac{1}{2} \\ 0 & 4 & -3 & | & 17 \end{bmatrix} \xrightarrow{R_3-4R_2} \begin{bmatrix} 1 & 0 & 1 & | & 4 \\ 0 & 1 & -\frac{1}{2} & | & \frac{1}{2} \\ 0 & 0 & -1 & | & 15 \end{bmatrix} \xrightarrow{-R_3} \begin{bmatrix} 1 & 0 & 1 & | & 4 \\ 0 & 1 & -\frac{1}{2} & | & \frac{1}{2} \\ 0 & 0 & 1 & | & -15 \end{bmatrix} \xrightarrow[R_2+\frac{1}{2}R_3]{R_1-R_3}$

$\begin{bmatrix} 1 & 0 & 0 & | & 19 \\ 0 & 1 & 0 & | & -7 \\ 0 & 0 & 1 & | & -15 \end{bmatrix}.$ The solution is $(19,-7,-15)$.

43. $\begin{bmatrix} 0 & -1 & 1 & | & 2 \\ 4 & -3 & 2 & | & 16 \\ 3 & 2 & 1 & | & 11 \end{bmatrix} \xrightarrow{R_1 \leftrightarrow R_2} \begin{bmatrix} 4 & -3 & 2 & | & 16 \\ 0 & -1 & 1 & | & 2 \\ 3 & 2 & 1 & | & 11 \end{bmatrix} \xrightarrow{R_1-R_3} \begin{bmatrix} 1 & -5 & 1 & | & 5 \\ 0 & -1 & 1 & | & 2 \\ 3 & 2 & 1 & | & 11 \end{bmatrix}$

$\xrightarrow[R_3-3R_1]{-R_2} \begin{bmatrix} 1 & -5 & 1 & | & 5 \\ 0 & 1 & -1 & | & -2 \\ 0 & 17 & -2 & | & -4 \end{bmatrix} \xrightarrow[R_3-17R_2]{R_1+5R_2} \begin{bmatrix} 1 & 0 & -4 & | & -5 \\ 0 & 1 & -1 & | & -2 \\ 0 & 0 & 15 & | & 30 \end{bmatrix} \xrightarrow{\frac{1}{15}R_3}$

$\begin{bmatrix} 1 & 0 & -4 & | & -5 \\ 0 & 1 & -1 & | & -2 \\ 0 & 0 & 1 & | & 2 \end{bmatrix} \xrightarrow[R_2+R_3]{R_1+4R_3} \begin{bmatrix} 1 & 0 & 0 & | & 3 \\ 0 & 1 & 0 & | & 0 \\ 0 & 0 & 1 & | & 2 \end{bmatrix}.$

The solution is $(3,0,2)$.

45. Using the Gauss-Jordan method, we have

$$\begin{bmatrix} 1 & -2 & 1 & | & 6 \\ 2 & 1 & -3 & | & -3 \\ 1 & -3 & 3 & | & 10 \end{bmatrix} \xrightarrow[R_3-R_1]{R_2-2R_1} \begin{bmatrix} 1 & -2 & 1 & | & 6 \\ 0 & 5 & -5 & | & -15 \\ 0 & -1 & 2 & | & 4 \end{bmatrix} \xrightarrow{\frac{1}{5}R_2} \begin{bmatrix} 1 & -2 & 1 & | & 6 \\ 0 & 1 & -1 & | & -3 \\ 0 & -1 & 2 & | & 4 \end{bmatrix}$$

$$\xrightarrow[R_3+R_2]{R_1+2R_2} \begin{bmatrix} 1 & 0 & -1 & | & 0 \\ 0 & 1 & -1 & | & -3 \\ 0 & 0 & 1 & | & 1 \end{bmatrix} \xrightarrow[R_2+R_3]{R_1+R_3} \begin{bmatrix} 1 & 0 & 0 & | & 1 \\ 0 & 1 & 0 & | & -2 \\ 0 & 0 & 1 & | & 1 \end{bmatrix}.$$

Therefore, the solution is $(1,-2,1)$.

47. Using the Gauss-Jordan method, we have

$$\begin{bmatrix} 2 & 0 & 3 & | & -1 \\ 3 & -2 & 1 & | & 9 \\ 1 & 1 & 4 & | & 4 \end{bmatrix} \xrightarrow{R_1 \leftrightarrow R_3} \begin{bmatrix} 1 & 1 & 4 & | & 4 \\ 3 & -2 & 1 & | & 9 \\ 2 & 0 & 3 & | & -1 \end{bmatrix} \xrightarrow[R_3-2R_1]{R_2-3R_1} \begin{bmatrix} 1 & 1 & 4 & | & 4 \\ 0 & -5 & -11 & | & -3 \\ 0 & -2 & -5 & | & -9 \end{bmatrix}$$

$$\xrightarrow{-\frac{1}{5}R_2} \begin{bmatrix} 1 & 1 & 4 & | & 4 \\ 0 & 1 & \frac{11}{5} & | & \frac{3}{5} \\ 0 & -2 & -5 & | & -9 \end{bmatrix} \xrightarrow[R_3+2R_2]{R_1-R_2} \begin{bmatrix} 1 & 0 & \frac{9}{5} & | & \frac{17}{5} \\ 0 & 1 & \frac{11}{5} & | & \frac{3}{5} \\ 0 & 0 & -\frac{3}{5} & | & -\frac{39}{5} \end{bmatrix} \xrightarrow{-\frac{5}{3}R_3}$$

$$\begin{bmatrix} 1 & 0 & \frac{9}{5} & | & \frac{17}{5} \\ 0 & 1 & \frac{11}{5} & | & \frac{3}{5} \\ 0 & 0 & 1 & | & 13 \end{bmatrix} \xrightarrow[R_2-\frac{11}{5}R_3]{R_1-\frac{9}{5}R_3} \begin{bmatrix} 1 & 0 & 0 & | & -20 \\ 0 & 1 & 0 & | & -28 \\ 0 & 0 & 1 & | & 13 \end{bmatrix}.$$

Therefore, the solution is $(-20, -28, 13)$.

49. Using the Gauss-Jordan method, we have

$$\begin{bmatrix} 1 & -1 & 3 & | & 14 \\ 1 & 1 & 1 & | & 6 \\ -2 & -1 & 1 & | & -4 \end{bmatrix} \xrightarrow[R_3+2R_1]{R_2-R_1} \begin{bmatrix} 1 & -1 & 3 & | & 14 \\ 0 & 2 & -2 & | & -8 \\ 0 & -3 & 7 & | & 24 \end{bmatrix} \xrightarrow{\frac{1}{2}R_2} \begin{bmatrix} 1 & -1 & 3 & | & 14 \\ 0 & 1 & -1 & | & -4 \\ 0 & -3 & 7 & | & 24 \end{bmatrix}$$

$$\xrightarrow[\substack{R_1+R_2 \\ R_3+3R_2}]{} \begin{bmatrix} 1 & 0 & 2 & | & 10 \\ 0 & 1 & -1 & | & -4 \\ 0 & 0 & 4 & | & 12 \end{bmatrix} \xrightarrow[\frac{1}{4}R_3]{} \begin{bmatrix} 1 & 0 & 2 & | & 10 \\ 0 & 1 & -1 & | & -4 \\ 0 & 0 & 1 & | & 3 \end{bmatrix} \xrightarrow[\substack{R_1-2R_3 \\ R_2+R_3}]{} \begin{bmatrix} 1 & 0 & 0 & | & 4 \\ 0 & 1 & 0 & | & -1 \\ 0 & 0 & 1 & | & 3 \end{bmatrix}$$

Therefore, the solution is $(4, -1, 3)$.

51. We wish to solve the system of equations

$$\begin{aligned} x + y &= 500 \\ 42x + 30y &= 18,600 \end{aligned}$$

$(x =$ the number of acres of corn planted)
$(y =$ the number of acres of wheat planted)

Using the Gauss-Jordan method, we find

$$\begin{bmatrix} 1 & 1 & | & 500 \\ 42 & 30 & | & 18600 \end{bmatrix} \xrightarrow[R_2-42R_1]{} \begin{bmatrix} 1 & 1 & | & 500 \\ 0 & -12 & | & -2400 \end{bmatrix} \xrightarrow[-\frac{1}{12}R_2]{} \begin{bmatrix} 1 & 1 & | & 500 \\ 0 & 1 & | & 200 \end{bmatrix}$$

$$\xrightarrow[R_1-R_2]{} \begin{bmatrix} 1 & 0 & | & 300 \\ 0 & 1 & | & 200 \end{bmatrix}.$$

The solution to this system of equations is $x = 300$ and $y = 200$. We conclude that Jacob should plant 300 acres of corn and 200 acres of wheat.

53. Let x denote the number of pounds of the $2.50/lb coffee and y denote the number of pounds of the $3.00/lb coffee. Then we are required to solve the system

$$\begin{aligned} x + y &= 100 \\ 2.50x + 3.00y &= 280 \end{aligned}$$

Using the Gauss-Jordan method of elimination, we have

$$\begin{bmatrix} 1 & 1 & | & 100 \\ 2.5 & 3 & | & 280 \end{bmatrix} \xrightarrow[R_2-2.5R_1]{} \begin{bmatrix} 1 & 1 & | & 100 \\ 0 & 0.5 & | & 30 \end{bmatrix} \xrightarrow[2R_2]{} \begin{bmatrix} 1 & 1 & | & 100 \\ 0 & 1 & | & 60 \end{bmatrix}$$

$$\xrightarrow[R_1-R_2]{} \begin{bmatrix} 1 & 0 & | & 40 \\ 0 & 1 & | & 60 \end{bmatrix}.$$

Therefore, 40 pounds of the $2.50/lb coffee and 60 pounds of the $3.00/lb coffee should be used in the 100 lb mixture.

55. Let x and y denote the number of children and adults who rode the bus during the morning shift, respectively. Then the solution to the problem can be found by solving the system of equations

$$x + y = 1000$$
$$0.25x + 0.75y = 650$$

Using the Gauss-Jordan elimination method, we have

$$\begin{bmatrix} 1 & 1 & | & 1000 \\ 0.25 & 0.75 & | & 650 \end{bmatrix} \xrightarrow{R_2 - 0.25R_1} \begin{bmatrix} 1 & 1 & | & 1000 \\ 0 & 0.5 & | & 400 \end{bmatrix} \xrightarrow{2R_2} \begin{bmatrix} 1 & 1 & | & 1000 \\ 0 & 1 & | & 800 \end{bmatrix}$$

$$\xrightarrow{R_1 - R_2} \begin{bmatrix} 1 & 0 & | & 200 \\ 0 & 1 & | & 800 \end{bmatrix}.$$

We conclude that 800 adults and 200 children rode the bus during the morning shift.

57. Let x, y, and z, denote the amount of money he should invest in a savings account, in mutual funds, and inbonds, respectively. Then, we are required to solve the system

$$0.06x + 0.08y + 0.12z = 21,600$$
$$2x \qquad - \quad z = 0$$
$$0.08y - 0.12z = 0$$

Using the Gauss-Jordan method, we find

$$\begin{bmatrix} 0.06 & 0.08 & 0.12 & | & 21,600 \\ 2 & 0 & -1 & | & 0 \\ 0 & 0.08 & -0.12 & | & 0 \end{bmatrix} \xrightarrow[\frac{1}{0.08}R_3]{\frac{1}{0.06}R_1} \begin{bmatrix} 1 & \frac{4}{3} & 2 & | & 360,000 \\ 2 & 0 & -1 & | & 0 \\ 0 & 1 & -\frac{3}{2} & | & 0 \end{bmatrix} \xrightarrow{R_2 - 2R_1}$$

$$\begin{bmatrix} 1 & \frac{4}{3} & 2 & | & 360,000 \\ 0 & -\frac{8}{3} & -5 & | & -720,000 \\ 0 & 1 & -\frac{3}{2} & | & 0 \end{bmatrix} \xrightarrow{-\frac{3}{8}R_2} \begin{bmatrix} 1 & \frac{4}{3} & 2 & | & 360,000 \\ 0 & 1 & \frac{15}{8} & | & 270,000 \\ 0 & 1 & -\frac{3}{2} & | & 0 \end{bmatrix}$$

2 Systems of Linear Equations and Matrices

$$\begin{array}{c} R_1 - \frac{4}{3}R_2 \\ R_3 - R_2 \\ \longrightarrow \end{array} \begin{bmatrix} 1 & 0 & -\frac{1}{2} & 0 \\ 0 & 1 & \frac{15}{8} & 270{,}000 \\ 0 & 0 & -\frac{27}{8} & -270{,}000 \end{bmatrix} \xrightarrow{-\frac{8}{27}R_3} \begin{bmatrix} 1 & 0 & -\frac{1}{2} & 0 \\ 0 & 1 & \frac{15}{8} & 270{,}000 \\ 0 & 0 & 1 & 80{,}000 \end{bmatrix}$$

$$\begin{array}{c} R_1 + \frac{1}{2}R_3 \\ R_2 - \frac{15}{8}R_3 \\ \longrightarrow \end{array} \begin{bmatrix} 1 & 0 & 0 & 40{,}000 \\ 0 & 1 & 0 & 120{,}000 \\ 0 & 0 & 1 & 80{,}000 \end{bmatrix}$$

Therefore, Sid should invest $40,000 in a savings account, $120,000 in mutual funds, and $80,000 in bonds.

59. Let x, y, and z denote the number of compact, intermediate, and full-size cars, respectively, to be purchased. Then the problem can be solved by solving the system

$$10000x + 15{,}000y + 20{,}000z = 1{,}250{,}000$$

$$x - \quad 2y \qquad = \qquad 0$$

$$x + \quad y + \quad z = \qquad 100$$

Using the Gauss-Jordan method, we have

$$\begin{bmatrix} 10{,}000 & 15{,}000 & 20{,}000 & 1{,}250{,}000 \\ 1 & -2 & 0 & 0 \\ 1 & 1 & 1 & 100 \end{bmatrix} \xrightarrow{R_1 \leftrightarrow R_3} \begin{bmatrix} 1 & 1 & 1 & 100 \\ 1 & -2 & 0 & 0 \\ 10000 & 15000 & 20000 & 1{,}250{,}000 \end{bmatrix}$$

$$\begin{array}{c} R_2 - R_1 \\ R_3 - 10{,}000R_1 \\ \longrightarrow \end{array} \begin{bmatrix} 1 & 1 & 1 & 100 \\ 0 & -3 & -1 & -100 \\ 0 & 5000 & 10000 & 250{,}000 \end{bmatrix} \xrightarrow{-\frac{1}{3}R_2} \begin{bmatrix} 1 & 1 & 1 & 100 \\ 0 & 1 & \frac{1}{3} & \frac{100}{3} \\ 0 & 5000 & 10000 & 250{,}000 \end{bmatrix}$$

$$\begin{array}{c} R_1 - R_2 \\ R_3 - 5000R_2 \\ \longrightarrow \end{array} \begin{bmatrix} 1 & 0 & \frac{2}{3} & \frac{200}{3} \\ 0 & 1 & \frac{1}{3} & \frac{100}{3} \\ 0 & 0 & \frac{25{,}000}{3} & \frac{250{,}000}{3} \end{bmatrix} \xrightarrow{\frac{3}{25{,}000}R_3} \begin{bmatrix} 1 & 0 & \frac{2}{3} & \frac{200}{3} \\ 0 & 1 & \frac{1}{3} & \frac{100}{3} \\ 0 & 0 & 1 & 10 \end{bmatrix} \begin{array}{c} R_1 - \frac{2}{3}R_3 \\ R_2 - \frac{1}{3}R_3 \\ \longrightarrow \end{array} \begin{bmatrix} 1 & 0 & 0 & 60 \\ 0 & 1 & 0 & 30 \\ 0 & 0 & 1 & 10 \end{bmatrix}.$$

We conclude that 60 compact cars, 30 intermediate-size cars, and 10 full-size cars will be purchased.

61. Let x, y, and z, represent the number of ounces of Food *I*, Food *II*, and Food *III* used in the meal, respectively. Then the problem reduces to solving the following system

of linear equations:

$$10x + 6y + 8z = 100$$
$$10x + 12y + 6z = 100$$
$$5x + 4y + 12z = 100.$$

Using the Gauss-Jordan method, we obtain

$$\begin{bmatrix} 10 & 6 & 8 & | & 100 \\ 10 & 12 & 6 & | & 100 \\ 5 & 4 & 12 & | & 100 \end{bmatrix} \xrightarrow{\frac{1}{10}R_1} \begin{bmatrix} 1 & \frac{3}{5} & \frac{4}{5} & | & 10 \\ 10 & 12 & 6 & | & 100 \\ 5 & 4 & 12 & | & 100 \end{bmatrix} \xrightarrow[R_3 - 5R_1]{R_2 - 10R_1}$$

$$\begin{bmatrix} 1 & \frac{3}{5} & \frac{4}{5} & | & 10 \\ 0 & 6 & -2 & | & 0 \\ 0 & 1 & 8 & | & 50 \end{bmatrix} \xrightarrow{\frac{1}{6}R_2} \begin{bmatrix} 1 & \frac{3}{5} & \frac{4}{5} & | & 10 \\ 0 & 1 & -\frac{1}{3} & | & 0 \\ 0 & 1 & 8 & | & 50 \end{bmatrix} \xrightarrow[R_3 - R_2]{R_1 - \frac{3}{5}R_2}$$

$$\begin{bmatrix} 1 & 0 & 1 & | & 10 \\ 0 & 1 & -\frac{1}{3} & | & 0 \\ 0 & 0 & \frac{25}{3} & | & 50 \end{bmatrix} \xrightarrow{\frac{3}{25}R_3} \begin{bmatrix} 1 & 0 & 1 & | & 10 \\ 0 & 1 & -\frac{1}{3} & | & 0 \\ 0 & 0 & 1 & | & 6 \end{bmatrix} \xrightarrow[R_2 + \frac{1}{3}R_3]{R_1 - R_3}$$

$$\begin{bmatrix} 1 & 0 & 0 & | & 4 \\ 0 & 1 & 0 & | & 2 \\ 0 & 0 & 1 & | & 6 \end{bmatrix}.$$

We conclude that 4 oz of Food *I*, 2 oz of Food *II*, and 6 oz of Food *III* should be used to prepare the meal.

63. Let x = the number of front orchestra seats sold
          y = the number of rear orchestra seats sold
    and z = the number of front balcony seats sold for this performance.
    Then, we are required to solve the system

$$x + y + z = 1{,}000$$
$$80x + 60y + 50z = 62{,}800$$
$$x + y - 2z = 400.$$

Using the Gauss-Jordan method, we find

$$\begin{bmatrix} 1 & 1 & 1 & 1,000 \\ 80 & 60 & 50 & 62,800 \\ 1 & 1 & -2 & 400 \end{bmatrix} \xrightarrow[R_3-R_1]{R_2-80R_1} \begin{bmatrix} 1 & 1 & 1 & 1,000 \\ 0 & -20 & -30 & -17,200 \\ 0 & 0 & -3 & -600 \end{bmatrix} \xrightarrow[-\frac{1}{3}R_3]{-\frac{1}{20}R_2}$$

$$\begin{bmatrix} 1 & 1 & 1 & 1,000 \\ 0 & 1 & \frac{3}{2} & 860 \\ 0 & 0 & 1 & 200 \end{bmatrix} \xrightarrow{R_1-R_2} \begin{bmatrix} 1 & 0 & -\frac{1}{2} & 140 \\ 0 & 1 & \frac{3}{2} & 860 \\ 0 & 0 & 1 & 200 \end{bmatrix} \xrightarrow[R_2-\frac{3}{2}R_3]{R_1+\frac{1}{2}R_3}$$

$$\begin{bmatrix} 1 & 0 & 0 & 240 \\ 0 & 1 & 0 & 560 \\ 0 & 0 & 1 & 200 \end{bmatrix}.$$

We conclude that tickets for 240 front orchestra seats, 560 rear orchestra seats, and 200 front balcony seats were sold.

65. False. The constant cannot be zero. The system

$$2x + y = 1$$
$$3x - y = 2$$

is not equivalent to

$$\begin{array}{c} 2x + y = 1 \\ 0(3x - y) = 0(2) \end{array} \quad \text{or} \quad \begin{array}{c} 2x + y = 1 \\ 0 = 0 \end{array}.$$

USING TECHNOLOGY EXERCISES 2.2, page 104

1. $(3,1,-1,2)$ 3. $(5,4,-3,-4)$ 5. $(1,-1,2,0,3)$

EXERCISES 2.3, page 113

1. a. The system has one solution.   b. The solution is $(3, -1, 2)$.

3. a. The system has one solution.   b. The solution is $(2, 4)$.

5. a. The system has infinitely many solutions.
   b. Letting $x_3 = t$, we see that the solutions are given by $(4 - t, -2, t)$, where t is a parameter.

7. a. The system has no solution. The last row contains all zeros to the left of the vertical line and a nonzero number (1) to its right.

9. a. The system has infinitely many solutions.
   b. Letting $x_4 = t$, we see that the solutions are given by $(2, -1, 2 - t, t)$, where t is a parameter.

11. a. The system has infinitely many solutions.
    b. Letting $x_3 = s$ and $x_4 = t$, the solutions are given by $(2 - 3s, 1 + s, s, t)$, where s and t are parameters.

13. Using the Gauss-Jordan method, we have

$$\begin{bmatrix} 2 & -1 & | & 3 \\ 1 & 2 & | & 4 \\ 2 & 3 & | & 7 \end{bmatrix} \xrightarrow{R_1 \leftrightarrow R_2} \begin{bmatrix} 1 & 2 & | & 4 \\ 2 & -1 & | & 3 \\ 2 & 3 & | & 7 \end{bmatrix} \xrightarrow[R_3-2R_1]{R_2-2R_1} \begin{bmatrix} 1 & 2 & | & 4 \\ 0 & -5 & | & -5 \\ 0 & -1 & | & -1 \end{bmatrix} \xrightarrow{-\frac{1}{5}R_2}$$

$$\begin{bmatrix} 1 & 2 & | & 4 \\ 0 & 1 & | & 1 \\ 0 & -1 & | & -1 \end{bmatrix} \xrightarrow[R_3+R_2]{R_1-2R_2} \begin{bmatrix} 1 & 0 & | & 2 \\ 0 & 1 & | & 1 \\ 0 & 0 & | & 0 \end{bmatrix}. \qquad \text{The solution is } (2,1).$$

15. Using the Gauss-Jordan method, we have

$$\begin{bmatrix} 3 & -2 & | & -3 \\ 2 & 1 & | & 3 \\ 1 & -2 & | & -5 \end{bmatrix} \xrightarrow{R_1 \leftrightarrow R_3} \begin{bmatrix} 1 & -2 & | & -5 \\ 2 & 1 & | & 3 \\ 3 & -2 & | & -3 \end{bmatrix} \xrightarrow[R_3-3R_1]{R_2-2R_1} \begin{bmatrix} 1 & -2 & | & -5 \\ 0 & 5 & | & 13 \\ 0 & 4 & | & 12 \end{bmatrix} \xrightarrow{\frac{1}{5}R_2}$$

$$\begin{bmatrix} 1 & -2 & | & -5 \\ 0 & 1 & | & \frac{13}{5} \\ 0 & 4 & | & 12 \end{bmatrix} \xrightarrow[R_3-4R_2]{R_1+2R_2} \begin{bmatrix} 1 & 0 & | & \frac{1}{5} \\ 0 & 1 & | & \frac{13}{5} \\ 0 & 0 & | & \frac{8}{5} \end{bmatrix}.$$

Since the last row implies the $0 = 8/5$, we conclude that the system of equations is inconsistent and has no solution.

17. $$\begin{bmatrix} 3 & -2 & | & 5 \\ -1 & 3 & | & -4 \\ 2 & -4 & | & 6 \end{bmatrix} \xrightarrow{R_1 \leftrightarrow R_2} \begin{bmatrix} -1 & 3 & | & -4 \\ 3 & -2 & | & 5 \\ 2 & -4 & | & 6 \end{bmatrix} \xrightarrow{-R_1} \begin{bmatrix} 1 & -3 & | & 4 \\ 3 & -2 & | & 5 \\ 2 & -4 & | & 6 \end{bmatrix} \xrightarrow[R_3-2R_1]{R_2-3R_1}$$

2 Systems of Linear Equations and Matrices

$$\begin{bmatrix} 1 & -3 & | & 4 \\ 0 & 7 & | & -7 \\ 0 & 2 & | & -2 \end{bmatrix} \xrightarrow{\frac{1}{7}R_2} \begin{bmatrix} 1 & -3 & | & 4 \\ 0 & 1 & | & -1 \\ 0 & 2 & | & -2 \end{bmatrix} \xrightarrow[R_3-2R_2]{R_1+3R_2} \begin{bmatrix} 1 & 0 & | & 1 \\ 0 & 1 & | & -1 \\ 0 & 0 & | & 0 \end{bmatrix}.$$

We conclude that the solution is $(1,-1)$.

19. $$\begin{bmatrix} 1 & -2 & | & 2 \\ 7 & -14 & | & 14 \\ 3 & -6 & | & 6 \end{bmatrix} \xrightarrow[R_3-3R_1]{R_2-7R_1} \begin{bmatrix} 1 & -2 & | & 2 \\ 0 & 0 & | & 0 \\ 0 & 0 & | & 0 \end{bmatrix}.$$

We conclude that the infinitely many solutions are given by $(2t + 2, t)$, where t is a parameter.

21. $$\begin{bmatrix} 3 & 2 & | & 4 \\ -\frac{3}{2} & -1 & | & -2 \\ 6 & 4 & | & 8 \end{bmatrix} \xrightarrow{\frac{1}{3}R_1} \begin{bmatrix} 1 & \frac{2}{3} & | & \frac{4}{3} \\ -\frac{3}{2} & -1 & | & -2 \\ 6 & 4 & | & 8 \end{bmatrix} \xrightarrow[R_3-6R_1]{R_2+\frac{3}{2}R_1} \begin{bmatrix} 1 & \frac{2}{3} & | & \frac{4}{3} \\ 0 & 0 & | & 0 \\ 0 & 0 & | & 0 \end{bmatrix}.$$

We conclude that the infinitely many solutions are given by $(\frac{4}{3}-\frac{2}{3}t, t)$, where t is a parameter.

23. $$\begin{bmatrix} 2 & -1 & 1 & | & -4 \\ 3 & -\frac{3}{2} & \frac{3}{2} & | & -6 \\ -6 & 3 & -3 & | & 12 \end{bmatrix} \xrightarrow{\frac{1}{2}R_1} \begin{bmatrix} 1 & -\frac{1}{2} & \frac{1}{2} & | & -2 \\ 3 & -\frac{3}{2} & \frac{3}{2} & | & -6 \\ -6 & 3 & -3 & | & 12 \end{bmatrix} \xrightarrow[R_3+6R_1]{R_2-3R_1} \begin{bmatrix} 1 & -\frac{1}{2} & \frac{1}{2} & | & -2 \\ 0 & 0 & 0 & | & 0 \\ 0 & 0 & 0 & | & 0 \end{bmatrix}.$$

We conclude that the infinitely many solutions are given by $(-2+\frac{1}{2}s-\frac{1}{2}t, s, t)$ where s and t are parameters.

25. $$\begin{bmatrix} 1 & -2 & 3 & | & 4 \\ 2 & 3 & -1 & | & 2 \\ 1 & 2 & -3 & | & -6 \end{bmatrix} \xrightarrow[R_3-R_1]{R_2-2R_1} \begin{bmatrix} 1 & -2 & 3 & | & 4 \\ 0 & 7 & -7 & | & -6 \\ 0 & 4 & -6 & | & -10 \end{bmatrix} \xrightarrow{\frac{1}{7}R_2} \begin{bmatrix} 1 & -2 & 3 & | & 4 \\ 0 & 1 & -1 & | & -\frac{6}{7} \\ 0 & 4 & -6 & | & -10 \end{bmatrix}$$

$$\xrightarrow[R_3-4R_2]{R_1+2R_2} \begin{bmatrix} 1 & 0 & 1 & | & \frac{16}{7} \\ 0 & 1 & -1 & | & -\frac{6}{7} \\ 0 & 0 & -2 & | & -\frac{46}{7} \end{bmatrix} \xrightarrow{-\frac{1}{2}R_3} \begin{bmatrix} 1 & 0 & 1 & | & \frac{16}{7} \\ 0 & 1 & -1 & | & -\frac{6}{7} \\ 0 & 0 & 1 & | & \frac{23}{7} \end{bmatrix} \xrightarrow[R_2+R_3]{R_1-R_3}$$

$$\begin{bmatrix} 1 & 0 & 0 & | & -1 \\ 0 & 1 & 0 & | & \frac{17}{7} \\ 0 & 0 & 1 & | & \frac{23}{7} \end{bmatrix}.$$

We conclude that the solution is $\left(-1, \frac{17}{7}, \frac{23}{7}\right)$.

27. $\begin{bmatrix} 4 & 1 & -1 & | & 4 \\ 8 & 2 & -2 & | & 8 \end{bmatrix} \xrightarrow{\frac{1}{4}R_1} \begin{bmatrix} 1 & \frac{1}{4} & \frac{1}{4} & | & 1 \\ 8 & 2 & -2 & | & 8 \end{bmatrix} \xrightarrow{R_2-8R_1} \begin{bmatrix} 1 & \frac{1}{4} & -\frac{1}{4} & | & 1 \\ 0 & 0 & 0 & | & 0 \end{bmatrix}$

We conclude that the infinitely many solutions are given by $\left(1 - \frac{1}{4}s + \frac{1}{4}t,\ s,\ t\right)$, where s and t are parameters.

29. $\begin{bmatrix} 2 & 1 & -3 & | & 1 \\ 1 & -1 & 2 & | & 1 \\ 5 & -2 & 3 & | & 6 \end{bmatrix} \xrightarrow{R_1 \leftrightarrow R_2} \begin{bmatrix} 1 & -1 & 2 & | & 1 \\ 2 & 1 & -3 & | & 1 \\ 5 & -2 & 3 & | & 6 \end{bmatrix} \xrightarrow[R_3-5R_1]{R_2-2R_1} \begin{bmatrix} 1 & -1 & 2 & | & 1 \\ 0 & 3 & -7 & | & -1 \\ 0 & 3 & -7 & | & 1 \end{bmatrix} \xrightarrow{\frac{1}{3}R_2}$

$\begin{bmatrix} 1 & -1 & 2 & | & 1 \\ 0 & 1 & -\frac{7}{3} & | & -\frac{1}{3} \\ 0 & 3 & -7 & | & 1 \end{bmatrix} \xrightarrow[R_3-3R_2]{R_1+R_2} \begin{bmatrix} 1 & 0 & -\frac{1}{3} & | & \frac{2}{3} \\ 0 & 1 & -\frac{7}{3} & | & -\frac{1}{3} \\ 0 & 0 & 0 & | & 2 \end{bmatrix}.$

This last row implies that $0 = 2$, which is impossible. We conclude that the system of equations is inconsistent and has no solution.

31. $\begin{bmatrix} 1 & 2 & -1 & | & -4 \\ 2 & 1 & 1 & | & 7 \\ 1 & 3 & 2 & | & 7 \\ 1 & -3 & 1 & | & 9 \end{bmatrix} \xrightarrow[\substack{R_3-R_1 \\ R_4-R_1}]{R_2-2R_1} \begin{bmatrix} 1 & 2 & -1 & | & -4 \\ 0 & -3 & 3 & | & 15 \\ 0 & 1 & 3 & | & 11 \\ 0 & -5 & 2 & | & 13 \end{bmatrix} \xrightarrow{-\frac{1}{3}R_2} \begin{bmatrix} 1 & 2 & -1 & | & -4 \\ 0 & 1 & -1 & | & -5 \\ 0 & 1 & 3 & | & 11 \\ 0 & -5 & 2 & | & 13 \end{bmatrix}$

$\xrightarrow[\substack{R_3-R_2 \\ R_4+5R_2}]{R_1-2R_2} \begin{bmatrix} 1 & 0 & 1 & | & 6 \\ 0 & 1 & -1 & | & -5 \\ 0 & 0 & 4 & | & 16 \\ 0 & 0 & -3 & | & -12 \end{bmatrix} \xrightarrow{\frac{1}{4}R_3} \begin{bmatrix} 1 & 0 & 1 & | & 6 \\ 0 & 1 & -1 & | & -5 \\ 0 & 0 & 1 & | & 4 \\ 0 & 0 & -3 & | & -12 \end{bmatrix} \xrightarrow[\substack{R_2+R_3 \\ R_4+3R_3}]{R_1+\frac{1}{3}R_3} \begin{bmatrix} 1 & 0 & 0 & | & 2 \\ 0 & 1 & 0 & | & -1 \\ 0 & 0 & 1 & | & 4 \\ 0 & 0 & 0 & | & 0 \end{bmatrix}.$

We conclude that the solution of the system is $(2, -1, 4)$.

33. Let x, y, and z represent the number of compact, mid-sized, and full-size cars, respectively, to be purchased. Then the problem can be solved by solving the system

$$\begin{aligned} x + \quad y + \quad z &= \quad 60 \\ 10000x + 16000y + 22000z &= 840000 \ . \end{aligned}$$

Using the Gauss-Jordan method, we have

$$\begin{bmatrix} 1 & 1 & 1 & | & 60 \\ 10000 & 16000 & 22000 & | & 840000 \end{bmatrix} \xrightarrow{R_2 - 10,000R_1} \begin{bmatrix} 1 & 1 & 1 & | & 60 \\ 0 & 6000 & 12000 & | & 240000 \end{bmatrix}$$

$$\xrightarrow{\frac{1}{6000}R_2} \begin{bmatrix} 1 & 1 & 1 & | & 60 \\ 0 & 1 & 2 & | & 40 \end{bmatrix} \xrightarrow{R_1 - R_2} \begin{bmatrix} 1 & 0 & -1 & | & 20 \\ 0 & 1 & 2 & | & 40 \end{bmatrix}$$

and we conclude that the solution is $(20 + z, 40 - 2z, z)$. Letting $z = 5$, we see that one possible solution is $(25,30,5)$; that is Hartman should buy 25 compact, 30 mid-sized cars, and 5 full-sized cars. Letting $z = 10$, we see that another possible solution is $(30,20,10)$; that is, 30 compact cars, 20 mid-sized cars, and 10 full-sized cars.

35. Let x, y, and z denote the number of ounces of Food I, Food II, and Food III, respectively, that the dietician includes in the meal. Then the problem can be solved by solving the system

$$\begin{aligned} 400x + 1200y + 800z &= 8800 \\ 110x + 570y + 340z &= 2160 \\ 90x + 30y + 60z &= 1020. \end{aligned}$$

Using the Gauss-Jordan method, we have

$$\begin{bmatrix} 400 & 1200 & 800 & | & 8800 \\ 110 & 570 & 340 & | & 2160 \\ 90 & 30 & 60 & | & 1020 \end{bmatrix} \xrightarrow{\frac{1}{400}R_1} \begin{bmatrix} 1 & 3 & 2 & | & 22 \\ 110 & 570 & 340 & | & 2160 \\ 90 & 30 & 60 & | & 1020 \end{bmatrix} \xrightarrow[R_3 - 90R_1]{R_2 - 110R_1}$$

$$\begin{bmatrix} 1 & 3 & 2 & | & 22 \\ 0 & 240 & 120 & | & -260 \\ 0 & -240 & -120 & | & -960 \end{bmatrix} \xrightarrow{\frac{1}{240}R_2} \begin{bmatrix} 1 & 3 & 2 & | & 22 \\ 0 & 1 & \frac{1}{2} & | & -\frac{13}{12} \\ 0 & -240 & -120 & | & -960 \end{bmatrix} \xrightarrow[R_3 + 240R_2]{R_1 - 3R_2}$$

$$\begin{bmatrix} 1 & 0 & \frac{1}{2} & | & \frac{101}{4} \\ 0 & 1 & \frac{1}{2} & | & -\frac{13}{12} \\ 0 & 0 & 0 & | & -1220 \end{bmatrix}.$$

This last row implies that $0 = -1220$, which is impossible. We conclude that the system of equations is inconsistent and has no solution--that is, the dietician cannot prepare a meal from these foods and meet the given requirements.

37. a.
$$
\begin{aligned}
x_1 - x_2 \qquad\qquad\qquad\qquad &= 200 \\
x_1 \qquad\qquad\quad - x_5 \qquad &= 100 \\
-x_2 + x_3 \qquad\qquad + x_6 &= 600 \\
-x_3 + x_4 \qquad\qquad &= 200 \\
x_4 - x_5 + x_6 &= 700.
\end{aligned}
$$

b.

$$
\begin{bmatrix}
1 & -1 & 0 & 0 & 0 & 0 & | & 200 \\
1 & 0 & 0 & 0 & -1 & 0 & | & 100 \\
0 & -1 & 1 & 0 & 0 & 1 & | & 600 \\
0 & 0 & -1 & 1 & 0 & 0 & | & 200 \\
0 & 0 & 0 & 1 & -1 & 1 & | & 700
\end{bmatrix}
\xrightarrow{R_2 - R_1}
\begin{bmatrix}
1 & -1 & 0 & 0 & 0 & 0 & | & 200 \\
0 & 1 & 0 & 0 & -1 & 0 & | & -100 \\
0 & -1 & 1 & 0 & 0 & 1 & | & 600 \\
0 & 0 & -1 & 1 & 0 & 0 & | & 200 \\
0 & 0 & 0 & 1 & -1 & 1 & | & 700
\end{bmatrix}
\xrightarrow[R_1 + R_2]{R_3 + R_2}
$$

$$
\begin{bmatrix}
1 & 0 & 0 & 0 & -1 & 0 & | & 100 \\
0 & 1 & 0 & 0 & -1 & 0 & | & -100 \\
0 & 0 & 1 & 0 & -1 & 1 & | & 500 \\
0 & 0 & -1 & 1 & 0 & 0 & | & 200 \\
0 & 0 & 0 & 1 & -1 & 1 & | & 700
\end{bmatrix}
\xrightarrow{R_4 + R_3}
\begin{bmatrix}
1 & 0 & 0 & 0 & -1 & 0 & | & 100 \\
0 & 1 & 0 & 0 & -1 & 0 & | & -100 \\
0 & 0 & 1 & 0 & -1 & 1 & | & 500 \\
0 & 0 & 0 & 1 & -1 & 1 & | & 700 \\
0 & 0 & 0 & 1 & -1 & 1 & | & 700
\end{bmatrix}
\xrightarrow{R_5 - R_4}
$$

$$
\begin{bmatrix}
1 & 0 & 0 & 0 & -1 & 0 & | & 100 \\
0 & 1 & 0 & 0 & 1 & 0 & | & -100 \\
0 & 0 & 1 & 0 & -1 & 1 & | & 500 \\
0 & 0 & 0 & 1 & -1 & 1 & | & 700 \\
0 & 0 & 0 & 0 & 0 & 0 & | & 0
\end{bmatrix}.
$$

We conclude that the solution is
$$(s+100,\ s-100,\ s-t+500,\ s-t+700,\ s,\ t).$$
Taking $s = 150$ and $t = 50$, we see that one possible traffic pattern is
$$(250, 50, 600, 800, 150, 50).$$
Similarly, taking $s = 200$, and $t = 100$, we see that another possible traffic pattern is
$$(300, 100, 600, 800, 200, 100).$$
c. Taking $t = 0$ and $s = 200$, we see that another possible traffic pattern is
$$(300, 100, 700, 900, 200, 0).$$

39. We solve the given system by using the Gauss-Jordan method. We have

$$\begin{bmatrix} 2 & 3 & | & 2 \\ 1 & 4 & | & 6 \\ 5 & k & | & 2 \end{bmatrix} \xrightarrow{R_1 \leftrightarrow R_2} \begin{bmatrix} 1 & 4 & | & 6 \\ 2 & 3 & | & 2 \\ 5 & k & | & 2 \end{bmatrix} \xrightarrow[R_3 - 4R_1]{R_2 - 2R_1} \begin{bmatrix} 1 & 4 & | & 6 \\ 0 & -5 & | & -10 \\ 0 & k-20 & | & -28 \end{bmatrix} \xrightarrow{-\frac{1}{5}R_2}$$

$$\begin{bmatrix} 1 & 4 & | & 6 \\ 0 & 1 & | & 2 \\ 0 & k-20 & | & -28 \end{bmatrix} \xrightarrow[R_3 + aR_2]{R_1 - 4R_2} \begin{bmatrix} 1 & 0 & | & -2 \\ 0 & 1 & | & 2 \\ 0 & k+a-20 & | & -28+2a \end{bmatrix}$$

From the last matrix, we see that the system has a solution if and only if $x = -2$, $y = 2$, and

$$-28 + 2a = 0, \text{ or } a = 14$$

and $\qquad k + a - 20 = k - 6 = 0, \text{ or } k = 6.$

(All the entries in the last row of the matrix must be equal to zero.)

41. False. Such a system cannot have a unique solution.

USING TECHNOLOGY EXERCISES 2.3, page 117

1. $(1+t, 2+t, t)$; t, a parameter   3. $\left(-\frac{17}{7} + \frac{6}{7}t, 3 - t, -\frac{18}{7} + \frac{1}{7}t, t\right)$ 5. No solution

EXERCISES 2.4, page 127

1. The size of A is 4×4; the size of B is 4×3; the size of C is 1×5, and the size of D is 4×1.

3. These are entries of the matrix B. The entry b_{13} refers to the entry in the first row and third column and is equal to 2. Similarly, $b_{31} = 3$, and $b_{43} = 8$.

5. The column matrix is the matrix D. The transpose of the matrix D is
$$D^T = [1 \ \ 3 \ \ -2 \ \ 0].$$

7. A is of size 3×2; B is of size 3×2; C and D are of size 3×3.

9.
$$A+B=\begin{bmatrix} -1 & 2 \\ 3 & -2 \\ 4 & 0 \end{bmatrix}+\begin{bmatrix} 2 & 4 \\ 3 & 1 \\ -2 & 2 \end{bmatrix}=\begin{bmatrix} 1 & 6 \\ 6 & -1 \\ 2 & 2 \end{bmatrix}.$$

11.
$$\begin{bmatrix} 3 & -1 & 0 \\ 2 & -2 & 3 \\ 4 & 6 & 2 \end{bmatrix}-\begin{bmatrix} 2 & -2 & 4 \\ 3 & 6 & 2 \\ -2 & 3 & 1 \end{bmatrix}=\begin{bmatrix} 1 & 1 & -4 \\ -1 & -8 & 1 \\ 6 & 3 & 1 \end{bmatrix}.$$

13.
$$\begin{bmatrix} 6 & 3 & 8 \\ 4 & 5 & 6 \end{bmatrix}-\begin{bmatrix} 3 & -2 & -1 \\ 0 & -5 & -7 \end{bmatrix}=\begin{bmatrix} 3 & 5 & 9 \\ 4 & 10 & 13 \end{bmatrix}.$$

15.
$$\begin{bmatrix} 1 & 4 & -5 \\ 3 & -8 & 6 \end{bmatrix}+\begin{bmatrix} 4 & 0 & -2 \\ 3 & 6 & 5 \end{bmatrix}-\begin{bmatrix} 2 & 8 & 9 \\ -11 & 2 & -5 \end{bmatrix}=\begin{bmatrix} 3 & -4 & -16 \\ 17 & -4 & 16 \end{bmatrix}.$$

17.
$$\begin{bmatrix} 1.2 & 4.5 & -4.2 \\ 8.2 & 6.3 & -3.2 \end{bmatrix}-\begin{bmatrix} 3.1 & 1.5 & -3.6 \\ 2.2 & -3.3 & -4.4 \end{bmatrix}=\begin{bmatrix} -1.9 & 3.0 & -0.6 \\ 6.0 & 9.6 & 1.2 \end{bmatrix}.$$

19. $\dfrac{1}{2}\begin{bmatrix} 1 & 0 & 0 & -4 \\ 3 & 0 & -1 & 6 \\ -2 & 1 & -4 & 2 \end{bmatrix}+\dfrac{4}{3}\begin{bmatrix} 3 & 0 & -1 & 4 \\ -2 & 1 & -6 & 2 \\ 8 & 2 & 0 & -2 \end{bmatrix}-\dfrac{1}{3}\begin{bmatrix} 3 & -9 & -1 & 0 \\ 6 & 2 & 0 & -6 \\ 0 & 1 & -3 & 1 \end{bmatrix}$

$$=\begin{bmatrix} \frac{7}{2} & 3 & -1 & \frac{10}{3} \\ -\frac{19}{6} & \frac{2}{3} & -\frac{17}{2} & \frac{23}{3} \\ \frac{29}{3} & \frac{17}{6} & -1 & -2 \end{bmatrix}.$$

21.
$$\begin{bmatrix} 2x-2 & 3 & 2 \\ 2 & 4 & y-2 \\ 2z & -3 & 2 \end{bmatrix}=\begin{bmatrix} 3 & u & 2 \\ 2 & 4 & 5 \\ 4 & -3 & 2 \end{bmatrix}.$$

Now, by the definition of equality of matrices,

2 Systems of Linear Equations and Matrices

$$u = 3$$

$2x - 2 = 3$ and $2x = 5$, or $x = 5/2$,

$y - 2 = 5$, and $y = 7$,

$2z = 4$, and $z = 2$.

23. $\begin{bmatrix} 1 & x \\ 2y & -3 \end{bmatrix} - 4\begin{bmatrix} 2 & -2 \\ 0 & 3 \end{bmatrix} = \begin{bmatrix} 3z & 10 \\ 4 & -u \end{bmatrix}$; $\begin{bmatrix} -7 & x+8 \\ 2y & -15 \end{bmatrix} = \begin{bmatrix} 3z & 10 \\ 4 & -u \end{bmatrix}$.

Now, by the definition of equality of matrices,

$-u = -15$, so $u = 15$

$x + 8 = 10$, so $x = 2$

$2y = 4$, so $y = 2$

$3z = -7$, so $z = -7/3$.

25. To verify the Commutative Law for matrix addition, let us show that $A + B = B + A$.

Now, $A + B = \begin{bmatrix} 2 & -4 & 3 \\ 4 & 2 & 1 \end{bmatrix} + \begin{bmatrix} 4 & -3 & 2 \\ 1 & 0 & 4 \end{bmatrix} = \begin{bmatrix} 6 & -7 & 5 \\ 5 & 2 & 5 \end{bmatrix}$

$= \begin{bmatrix} 4 & -3 & 2 \\ 1 & 0 & 4 \end{bmatrix} + \begin{bmatrix} 2 & -4 & 3 \\ 4 & 2 & 1 \end{bmatrix} = B + A$.

27. $(3+5)A = 8A = 8\begin{bmatrix} 3 & 1 \\ 2 & 4 \\ -4 & 0 \end{bmatrix} = \begin{bmatrix} 24 & 8 \\ 16 & 32 \\ -32 & 0 \end{bmatrix} = 3\begin{bmatrix} 3 & 1 \\ 2 & 4 \\ -4 & 0 \end{bmatrix} + 5\begin{bmatrix} 3 & 1 \\ 2 & 4 \\ -4 & 0 \end{bmatrix}$

$= 3A + 5A$.

29. $4(A+B) = 4\left(\begin{bmatrix} 3 & 1 \\ 2 & 4 \\ -4 & 0 \end{bmatrix} + \begin{bmatrix} 1 & 2 \\ -1 & 0 \\ 3 & 2 \end{bmatrix} \right) = 4\begin{bmatrix} 4 & 3 \\ 1 & 4 \\ -1 & 2 \end{bmatrix} = \begin{bmatrix} 16 & 12 \\ 4 & 16 \\ -4 & 8 \end{bmatrix}$

$4A + 4B = 4\begin{bmatrix} 3 & 1 \\ 2 & 4 \\ -4 & 0 \end{bmatrix} + 4\begin{bmatrix} 1 & 2 \\ -1 & 0 \\ 3 & 2 \end{bmatrix} = \begin{bmatrix} 16 & 12 \\ 4 & 16 \\ -4 & 8 \end{bmatrix}$.

31. $[3 \quad 2 \quad -1 \quad 5]^T = \begin{bmatrix} 3 \\ 2 \\ -1 \\ 5 \end{bmatrix}.$

33. $\begin{bmatrix} 1 & -1 & 2 \\ 3 & 4 & 2 \\ 0 & 1 & 0 \end{bmatrix}^T = \begin{bmatrix} 1 & 3 & 0 \\ -1 & 4 & 1 \\ 2 & 2 & 0 \end{bmatrix}.$

35.

$\begin{array}{c} \\ Mr.\ Cross \\ Mr.\ Jones \\ Mr.\ Smith \end{array} \begin{array}{cccc} 1 & 2 & 3 & 4 \\ \begin{bmatrix} 220 & 215 & 210 & 205 \\ 220 & 210 & 200 & 195 \\ 215 & 205 & 195 & 190 \end{bmatrix} \end{array}$

37. a. $D = A + B - C$

$= \begin{bmatrix} 2820 & 1470 & 1120 \\ 1030 & 520 & 480 \\ 1170 & 540 & 460 \end{bmatrix} + \begin{bmatrix} 260 & 120 & 110 \\ 140 & 60 & 50 \\ 120 & 70 & 50 \end{bmatrix} - \begin{bmatrix} 120 & 80 & 80 \\ 70 & 30 & 40 \\ 60 & 20 & 40 \end{bmatrix}$

$= \begin{bmatrix} 2960 & 1510 & 1150 \\ 1100 & 550 & 490 \\ 1230 & 590 & 470 \end{bmatrix}.$

b. $E = 1.1D = 1.1 \begin{bmatrix} 2960 & 1510 & 1150 \\ 1100 & 550 & 490 \\ 1230 & 590 & 470 \end{bmatrix} = \begin{bmatrix} 3256 & 1661 & 1265 \\ 1210 & 605 & 539 \\ 1353 & 649 & 517 \end{bmatrix}.$

39. False. Take $\begin{bmatrix} 1 & 2 \\ 3 & 4 \end{bmatrix}$ and $c = 2$. Then

$cA = 2 \begin{bmatrix} 1 & 2 \\ 3 & 4 \end{bmatrix} = \begin{bmatrix} 2 & 4 \\ 6 & 8 \end{bmatrix}$ and $(cA)^T = \begin{bmatrix} 2 & 6 \\ 4 & 8 \end{bmatrix}.$

On the other hand, $\dfrac{1}{c}A^T = \dfrac{1}{2}\begin{bmatrix} 1 & 3 \\ 2 & 4 \end{bmatrix} = \begin{bmatrix} \frac{1}{2} & \frac{3}{2} \\ 1 & 2 \end{bmatrix} \neq (cA)^T$

1. $\begin{bmatrix} 15 & 38.75 & -67.5 & 33.75 \\ 51.25 & 40 & 52.5 & -38.75 \\ 21.25 & 35 & -65 & 105 \end{bmatrix}$

3. $\begin{bmatrix} -5 & 6.3 & -6.8 & 3.9 \\ 1 & 0.5 & 5.4 & -4.8 \\ 0.5 & 4.2 & -3.5 & 5.6 \end{bmatrix}$

5. $\begin{bmatrix} 16.44 & -3.65 & -3.66 & 0.63 \\ 12.77 & 10.64 & 2.58 & 0.05 \\ 5.09 & 0.28 & -10.84 & 17.64 \end{bmatrix}$

7. $\begin{bmatrix} 7.4 & 7.2 & 2.9 \\ -0.1 & 5.9 & 1.4 \\ -4 & 3 & -6.9 \\ 1.5 & -1.4 & 11.2 \end{bmatrix}$

EXERCISES 2.5, page 140

1. $(2 \times 3)(3 \times 5)$ so AB has order 2×5.
      ↑ ↑
       =

   $(3 \times 5)(2 \times 3)$ so BA is not defined.
      ↑ ↑
       ≠

3. $(1 \times 7)\,(7 \times 1)$ so AB has order 1×1.
      ↑ ↑
       =

   $(7 \times 1)\,(1 \times 7)$ so AB has order 7×7.
      ↑ ↑
       =

5. If AB and BA are defined then $n = s$ and $m = t$.

7. $\begin{bmatrix} 1 & 2 \\ 3 & 0 \end{bmatrix}\begin{bmatrix} 1 \\ -1 \end{bmatrix} = \begin{bmatrix} -1 \\ 3 \end{bmatrix}$

9. $\begin{bmatrix} 3 & 1 & 2 \\ -1 & 2 & 4 \end{bmatrix}\begin{bmatrix} 4 \\ 1 \\ -2 \end{bmatrix} = \begin{bmatrix} 9 \\ -10 \end{bmatrix}$

11. $\begin{bmatrix} -1 & 2 \\ 3 & 1 \end{bmatrix}\begin{bmatrix} 2 & 4 \\ 3 & 1 \end{bmatrix} = \begin{bmatrix} 4 & -2 \\ 9 & 13 \end{bmatrix}$

13. $\begin{bmatrix} 2 & 1 & 2 \\ 3 & 2 & 4 \end{bmatrix}\begin{bmatrix} -1 & 2 \\ 4 & 3 \\ 0 & 1 \end{bmatrix} = \begin{bmatrix} 2 & 9 \\ 5 & 16 \end{bmatrix}$

15.
$$\begin{bmatrix} 0.1 & 0.9 \\ 0.2 & 0.8 \end{bmatrix}\begin{bmatrix} 1.2 & 0.4 \\ 0.5 & 2.1 \end{bmatrix} = \begin{bmatrix} 0.1(1.2)+0.9(0.5) & 0.1(0.4)+0.9(2.1) \\ 0.2(1.2)+0.8(0.5) & 0.2(0.4)+0.8(2.1) \end{bmatrix}$$

$$= \begin{bmatrix} 0.57 & 1.93 \\ 0.64 & 1.76 \end{bmatrix}.$$

17.
$$\begin{bmatrix} 6 & -3 & 0 \\ -2 & 1 & -8 \\ 4 & -4 & 9 \end{bmatrix}\begin{bmatrix} 1 & 0 & 0 \\ 0 & 1 & 0 \\ 0 & 0 & 1 \end{bmatrix} = \begin{bmatrix} 6 & -3 & 0 \\ -2 & 1 & -8 \\ 4 & -4 & 9 \end{bmatrix}.$$

19.
$$\begin{bmatrix} 3 & 0 & -2 & 1 \\ 1 & 2 & 0 & -1 \end{bmatrix}\begin{bmatrix} 2 & 1 & -1 \\ -1 & 2 & 0 \\ 0 & 0 & 1 \\ -1 & -2 & 2 \end{bmatrix} = \begin{bmatrix} 5 & 1 & -3 \\ 1 & 7 & -3 \end{bmatrix}.$$

21.
$$4\begin{bmatrix} 1 & -2 & 0 \\ 2 & -1 & 1 \\ 3 & 0 & -1 \end{bmatrix}\begin{bmatrix} 1 & 3 & 1 \\ 1 & 4 & 0 \\ 0 & 1 & -2 \end{bmatrix} = \begin{bmatrix} -4 & -20 & 4 \\ 4 & 12 & 0 \\ 12 & 32 & 20 \end{bmatrix}$$

23.
$$\begin{bmatrix} 1 & 0 \\ 0 & 1 \end{bmatrix}\begin{bmatrix} 4 & -3 & 2 \\ 7 & 1 & -5 \end{bmatrix}\begin{bmatrix} 1 & 0 & 0 \\ 0 & 1 & 0 \\ 0 & 0 & 1 \end{bmatrix} = \begin{bmatrix} 1 & 0 \\ 0 & 1 \end{bmatrix}\begin{bmatrix} 4 & -3 & 2 \\ 7 & 1 & -5 \end{bmatrix} = \begin{bmatrix} 4 & -3 & 2 \\ 7 & 1 & -5 \end{bmatrix}.$$

25. To verify the associative law for matrix multiplication, we will show that $(AB)C = A(BC)$.

$$AB = \begin{bmatrix} 1 & 0 & -2 \\ 1 & -3 & 2 \\ -2 & 1 & 1 \end{bmatrix}\begin{bmatrix} 3 & 1 & 0 \\ 2 & 2 & 0 \\ 1 & -3 & -1 \end{bmatrix} = \begin{bmatrix} 1 & 7 & 2 \\ -1 & -11 & -2 \\ -3 & -3 & -1 \end{bmatrix}$$

$$(AB)C = \begin{bmatrix} 1 & 7 & 2 \\ -1 & -11 & -2 \\ -3 & -3 & -1 \end{bmatrix}\begin{bmatrix} 2 & -1 & 0 \\ 1 & -1 & 2 \\ 3 & -2 & 1 \end{bmatrix} = \begin{bmatrix} 15 & -12 & 16 \\ -19 & 16 & -24 \\ -12 & 8 & -7 \end{bmatrix}$$

$$BC = \begin{bmatrix} 3 & 1 & 0 \\ 2 & 2 & 0 \\ 1 & -3 & -1 \end{bmatrix} \begin{bmatrix} 2 & -1 & 0 \\ 1 & -1 & 2 \\ 3 & -2 & 1 \end{bmatrix} = \begin{bmatrix} 7 & -4 & 2 \\ 6 & -4 & 4 \\ -4 & 4 & -7 \end{bmatrix}$$

$$A(BC) = \begin{bmatrix} 1 & 0 & -2 \\ 1 & -3 & 2 \\ -2 & 1 & 1 \end{bmatrix} \begin{bmatrix} 7 & -4 & 2 \\ 6 & -4 & 4 \\ -4 & 4 & -7 \end{bmatrix} = \begin{bmatrix} 15 & -12 & 16 \\ -19 & 16 & -24 \\ -12 & 8 & -7 \end{bmatrix}.$$

27. $$AB = \begin{bmatrix} 1 & 2 \\ 3 & 4 \end{bmatrix} \begin{bmatrix} 2 & 1 \\ 4 & 3 \end{bmatrix} = \begin{bmatrix} 10 & 7 \\ 22 & 15 \end{bmatrix}$$

$$BA = \begin{bmatrix} 2 & 1 \\ 4 & 3 \end{bmatrix} \begin{bmatrix} 1 & 2 \\ 3 & 4 \end{bmatrix} = \begin{bmatrix} 5 & 8 \\ 13 & 20 \end{bmatrix}$$

Therefore, $AB \neq BA$ and matrix multiplication is not commutative.

29. $$AB = \begin{bmatrix} 3 & 0 \\ 8 & 0 \end{bmatrix} \begin{bmatrix} 0 & 0 \\ 4 & 5 \end{bmatrix} = \begin{bmatrix} 0 & 0 \\ 0 & 0 \end{bmatrix}$$

$AB = 0$, but neither A nor B is the zero matrix. Therefore, $AB = 0$, does not imply that A or B is the zero matrix.

31. $$\begin{bmatrix} a & b \\ c & d \end{bmatrix} \begin{bmatrix} 1 & 0 \\ -1 & 3 \end{bmatrix} = \begin{bmatrix} a-b & 3b \\ c-d & 3d \end{bmatrix} = \begin{bmatrix} -1 & -3 \\ 3 & 6 \end{bmatrix}$$

Then $3b = -3,$ and $b = -1$
      $3d = 6,$ and $d = 2$
      $a - b = -1,$ and $a = b - 1 = -2.$
      $c - d = 3,$ and $c = d + 3 = 5$

Therefore, $A = \begin{bmatrix} -2 & -1 \\ 5 & 2 \end{bmatrix}.$

33. a.

$$A^T = \begin{bmatrix} 2 & 5 \\ 4 & -6 \end{bmatrix} \text{ and } (A^T)^T = \begin{bmatrix} 2 & 4 \\ 5 & -6 \end{bmatrix} = A$$

b. $(A+B)^T = \begin{bmatrix} 6 & 12 \\ -2 & -3 \end{bmatrix}^T = \begin{bmatrix} 6 & -2 \\ 12 & -3 \end{bmatrix}$

$A^T + B^T = \begin{bmatrix} 2 & 5 \\ 4 & -6 \end{bmatrix} + \begin{bmatrix} 4 & -7 \\ 8 & 3 \end{bmatrix} = \begin{bmatrix} 6 & -2 \\ 12 & -3 \end{bmatrix}$

c. $AB = \begin{bmatrix} 2 & 4 \\ 5 & -6 \end{bmatrix} \begin{bmatrix} 4 & 8 \\ -7 & 3 \end{bmatrix} = \begin{bmatrix} -20 & 28 \\ 62 & 22 \end{bmatrix}$

so $(AB)^T = \begin{bmatrix} -20 & 62 \\ 28 & 22 \end{bmatrix}$.

$B^T A^T = \begin{bmatrix} 4 & -7 \\ 8 & 3 \end{bmatrix} \begin{bmatrix} 2 & 5 \\ 4 & -6 \end{bmatrix} = \begin{bmatrix} -20 & 62 \\ 28 & 22 \end{bmatrix} = (AB)^T$

35. The given system of linear equations can be represented by the matrix equation $AX = B$, where

$A = \begin{bmatrix} 2 & -3 \\ 3 & -4 \end{bmatrix}$, $X = \begin{bmatrix} x \\ y \end{bmatrix}$, and $B = \begin{bmatrix} 7 \\ 8 \end{bmatrix}$.

37. The given system of linear equations can be represented by the matrix equation $AX = B$, where

$A = \begin{bmatrix} 2 & -3 & 4 \\ 0 & 2 & -3 \\ 1 & -1 & 2 \end{bmatrix}$, $X = \begin{bmatrix} x \\ y \\ z \end{bmatrix}$, $B = \begin{bmatrix} 6 \\ 7 \\ 4 \end{bmatrix}$.

39. The given system of linear equations can be represented by the matrix equation $AX = B$, where

$A = \begin{bmatrix} -1 & 1 & 1 \\ 2 & -1 & -1 \\ -3 & 2 & 4 \end{bmatrix}$, $X = \begin{bmatrix} x_1 \\ x_2 \\ x_3 \end{bmatrix}$, $B = \begin{bmatrix} 0 \\ 2 \\ 4 \end{bmatrix}$.

41.   a. $AB = \begin{bmatrix} 200 & 300 & 100 & 200 \\ 100 & 200 & 400 & 0 \end{bmatrix} \begin{bmatrix} 54 \\ 48 \\ 98 \\ 82 \end{bmatrix} = \begin{bmatrix} 51,400 \\ 54,200 \end{bmatrix}$

b. The first entry shows that William's total stock holdings are $51,400, while Michael's stockholdings are $54,200.

43.   The column vector that represents the profit for each type of house is

$$B = \begin{bmatrix} 20,000 \\ 22,000 \\ 25,000 \\ 30,000 \end{bmatrix}.$$

The column vector that gives the total profit for Bond Brothers is

$$AB = \begin{bmatrix} 60 & 80 & 120 & 40 \\ 20 & 30 & 60 & 10 \\ 10 & 15 & 30 & 5 \end{bmatrix} \begin{bmatrix} 20,000 \\ 22,000 \\ 25,000 \\ 30,000 \end{bmatrix}$$

$$= \begin{bmatrix} 7,160,000 \\ 2,860,000 \\ 1,430,000 \end{bmatrix}.$$

Therefore, Bond Brothers expects to make $7,160,000 in New York, $2,860,000 in Connecticut, and $1,430,000 in Massachusetts, and the total profit is $11,450,000.

45.

$$\begin{array}{ccc} & D & R & I \end{array}$$
$$BA = \begin{bmatrix} 30,000 & 40,000 & 20,000 \end{bmatrix} \begin{bmatrix} 0.50 & 0.30 & 0.20 \\ 0.45 & 0.40 & 0.15 \\ 0.40 & 0.50 & 0.10 \end{bmatrix}$$

$$\begin{array}{ccc} D & R & I \end{array}$$
$$= \begin{bmatrix} 41,000 & 35,000 & 14,000 \end{bmatrix}$$

47. $AB = \begin{bmatrix} 2700 & 3000 \\ 800 & 700 \\ 500 & 300 \end{bmatrix} \begin{bmatrix} 0.25 & 0.20 & 0.30 & 0.25 \\ 0.30 & 0.35 & 0.25 & 0.10 \end{bmatrix} = \begin{bmatrix} 1575 & 1590 & 1560 & 975 \\ 410 & 405 & 415 & 270 \\ 215 & 205 & 225 & 155 \end{bmatrix}$

49. a. $AC = \begin{bmatrix} 320 & 280 & 460 & 280 \\ 480 & 360 & 580 & 0 \\ 540 & 420 & 200 & 880 \end{bmatrix} \begin{bmatrix} 120 \\ 180 \\ 260 \\ 500 \end{bmatrix} = \begin{bmatrix} 348,400 \\ 273,200 \\ 632,400 \end{bmatrix}$.

The entries give the total production costs at locations I, II, and III for the month of May as \$348,400, \$273,200, and \$632,400, respectively.

b. $AD = \begin{bmatrix} 320 & 280 & 460 & 280 \\ 480 & 360 & 580 & 0 \\ 540 & 420 & 200 & 880 \end{bmatrix} \begin{bmatrix} 160 \\ 250 \\ 350 \\ 700 \end{bmatrix} = \begin{bmatrix} 478,200 \\ 369,800 \\ 877,400 \end{bmatrix}$

The total revenue realized at locations I, II, and III for the month of May are \$478,200, \$369,800, and \$877,400, respectively.

c. $BC = \begin{bmatrix} 210 & 180 & 330 & 180 \\ 400 & 300 & 450 & 40 \\ 420 & 280 & 180 & 740 \end{bmatrix} \begin{bmatrix} 120 \\ 180 \\ 260 \\ 500 \end{bmatrix} = \begin{bmatrix} 233,400 \\ 239,000 \\ 517,600 \end{bmatrix}$.

The total production costs at locations I, II, and III for the month of June are \$233,400, \$239,000, and \$517,600, respectively.

d. $BD = \begin{bmatrix} 210 & 180 & 330 & 180 \\ 400 & 300 & 450 & 40 \\ 420 & 280 & 180 & 740 \end{bmatrix} \begin{bmatrix} 160 \\ 250 \\ 350 \\ 700 \end{bmatrix} = \begin{bmatrix} 320,100 \\ 324,500 \\ 718,200 \end{bmatrix}$.

The total revenue realized at locations I, II, and III for the month of June are \$320,100, \$324,500, and \$718,200, respectively.

2 Systems of Linear Equations and Matrices

e. $(A+B)C = \begin{bmatrix} 530 & 460 & 790 & 460 \\ 880 & 660 & 1030 & 40 \\ 960 & 700 & 380 & 1620 \end{bmatrix} \begin{bmatrix} 120 \\ 180 \\ 260 \\ 500 \end{bmatrix} = \begin{bmatrix} 581,800 \\ 512,200 \\ 1,150,000 \end{bmatrix}$.

The total production costs in May and June are Locations I, II, and III are $581,800, $512,200, and $1,150,000, respectively.

f. $(A+B)D = \begin{bmatrix} 530 & 460 & 790 & 460 \\ 880 & 660 & 1030 & 40 \\ 960 & 700 & 380 & 1620 \end{bmatrix} \begin{bmatrix} 160 \\ 250 \\ 350 \\ 700 \end{bmatrix} = \begin{bmatrix} 798,300 \\ 694,300 \\ 1,595,600 \end{bmatrix}$.

The total revenue realized in May and June in Locations I, II, and III are $798,300, $694,300, and $1,595,600, respectively.

g. $A(D-C) = \begin{bmatrix} 320 & 280 & 460 & 280 \\ 480 & 360 & 580 & 0 \\ 540 & 420 & 200 & 880 \end{bmatrix} \begin{bmatrix} 40 \\ 70 \\ 90 \\ 200 \end{bmatrix} = \begin{bmatrix} 129,800 \\ 96,600 \\ 245,000 \end{bmatrix}$.

The profits in Locations I, II, and III in May are $129,800, $96,600, and $245,000, respectively.

h. $B(D-C) = \begin{bmatrix} 210 & 180 & 330 & 180 \\ 400 & 300 & 450 & 40 \\ 420 & 280 & 180 & 740 \end{bmatrix} \begin{bmatrix} 40 \\ 70 \\ 90 \\ 200 \end{bmatrix} = \begin{bmatrix} 86,700 \\ 85,500 \\ 200,600 \end{bmatrix}$.

The profits in Locations I, II, and III in June are $86,700, $85,500, and $200,600, respectively.

i. $(A+B)(D-C) = \begin{bmatrix} 530 & 460 & 790 & 460 \\ 880 & 660 & 1030 & 40 \\ 960 & 700 & 380 & 1620 \end{bmatrix} \begin{bmatrix} 40 \\ 70 \\ 90 \\ 200 \end{bmatrix} = \begin{bmatrix} 216,500 \\ 182,100 \\ 445,600 \end{bmatrix}$.

The profits in Locations I, II, and III in May and June are $216,500, $182,100, $445,600, respectively.

51. False. Let A be a matrix of order 2×3 and let B be a matrix of order 3×2. Then AB and BA are both defined. But, evidently, neither A nor B is a square matrix.

USING TECHNOLOGY EXERCISES 2.5, page 137

1.
$$\begin{bmatrix} 18.66 & 15.2 & -12 \\ 24.48 & 41.88 & 89.82 \\ 15.39 & 7.16 & -1.25 \end{bmatrix}$$

3.
$$\begin{bmatrix} 20.09 & 20.61 & -1.3 \\ 44.42 & 71.6 & 64.89 \\ 20.97 & 7.17 & -60.65 \end{bmatrix}$$

5.
$$\begin{bmatrix} 32.89 & 13.63 & -57.17 \\ -12.85 & -8.37 & 256.92 \\ 13.48 & 14.29 & 181.64 \end{bmatrix}$$

7.
$$\begin{bmatrix} 18.66 & 24.48 & 15.39 \\ 15.2 & 41.88 & 7.16 \\ -12 & 89.82 & -1.25 \end{bmatrix}$$

9.
$$\begin{bmatrix} 87 & 68 & 110 & 82 \\ 119 & 176 & 221 & 143 \\ 51 & 128 & 142 & 94 \\ 28 & 174 & 174 & 112 \end{bmatrix} \quad \begin{bmatrix} 113 & 117 & 72 & 101 & 90 \\ 72 & 85 & 36 & 72 & 76 \\ 81 & 69 & 76 & 87 & 30 \\ 133 & 157 & 56 & 121 & 146 \\ 154 & 157 & 94 & 127 & 122 \end{bmatrix}$$

11.
$$\begin{bmatrix} 170 & 18.1 & 133.1 & -106.3 & 341.3 \\ 349 & 226.5 & 324.1 & 164 & 506.4 \\ 245.2 & 157.7 & 231.5 & 125.5 & 312.9 \\ 310 & 245.2 & 291 & 274.3 & 354.2 \end{bmatrix}$$

EXERCISES 2.6, page 158

1.
$$\begin{bmatrix} 1 & -3 \\ 1 & -2 \end{bmatrix}\begin{bmatrix} -2 & 3 \\ -1 & 1 \end{bmatrix} = \begin{bmatrix} 1 & 0 \\ 0 & 1 \end{bmatrix}; \quad \begin{bmatrix} -2 & 3 \\ -1 & 1 \end{bmatrix}\begin{bmatrix} 1 & -3 \\ 1 & -2 \end{bmatrix} = \begin{bmatrix} 1 & 0 \\ 0 & 1 \end{bmatrix}$$

3.
$$\begin{bmatrix} 3 & 2 & 3 \\ 2 & 2 & 1 \\ 2 & 1 & 1 \end{bmatrix}\begin{bmatrix} -\frac{1}{3} & -\frac{1}{3} & \frac{4}{3} \\ 0 & 1 & -1 \\ \frac{2}{3} & -\frac{1}{3} & -\frac{2}{3} \end{bmatrix} = \begin{bmatrix} 1 & 0 & 0 \\ 0 & 1 & 0 \\ 0 & 0 & 1 \end{bmatrix} \quad \text{and}$$

$$\begin{bmatrix} -\frac{1}{3} & -\frac{1}{3} & \frac{4}{3} \\ 0 & 1 & -1 \\ \frac{2}{3} & -\frac{1}{3} & -\frac{2}{3} \end{bmatrix} \begin{bmatrix} 3 & 2 & 3 \\ 2 & 2 & 1 \\ 2 & 1 & 1 \end{bmatrix} = \begin{bmatrix} 1 & 0 & 0 \\ 0 & 1 & 0 \\ 0 & 0 & 1 \end{bmatrix}.$$

5. Using Formula (13), we find

$$A^{-1} = \frac{1}{(2)(3)-(1)(5)}\begin{bmatrix} 3 & -5 \\ -1 & 2 \end{bmatrix} = \begin{bmatrix} 3 & -5 \\ -1 & 2 \end{bmatrix}.$$

7. Since $ad - bc = (3)(2) - (-2)(-3) = 6 - 6 = 0$, the inverse does not exist.

9.
$$\begin{bmatrix} 2 & -3 & -4 & | & 1 & 0 & 0 \\ 0 & 0 & -1 & | & 0 & 1 & 0 \\ 1 & -2 & 1 & | & 0 & 0 & 1 \end{bmatrix} \xrightarrow{R_1 \leftrightarrow R_3} \begin{bmatrix} 1 & -2 & 1 & | & 0 & 0 & 1 \\ 0 & 0 & -1 & | & 0 & 1 & 0 \\ 2 & -3 & -4 & | & 1 & 0 & 0 \end{bmatrix} \xrightarrow{R_3 - 2R_1}$$

$$\begin{bmatrix} 1 & -2 & 1 & | & 0 & 0 & 1 \\ 0 & 0 & -1 & | & 0 & 1 & 0 \\ 0 & 1 & -6 & | & 1 & 0 & -2 \end{bmatrix} \xrightarrow{R_2 \leftrightarrow R_3} \begin{bmatrix} 1 & -2 & 1 & | & 0 & 0 & 1 \\ 0 & 1 & -6 & | & 1 & 0 & -2 \\ 0 & 0 & -1 & | & 0 & 1 & 0 \end{bmatrix} \xrightarrow[\;-R_3\;]{R_1 + 2R_2}$$

$$\begin{bmatrix} 1 & 0 & -11 & | & 2 & 0 & -3 \\ 0 & 1 & -6 & | & 1 & 0 & -2 \\ 0 & 0 & 1 & | & 0 & -1 & 0 \end{bmatrix} \xrightarrow[R_2 + 6R_3]{R_1 + 11R_3} \begin{bmatrix} 1 & 0 & 0 & | & 2 & -11 & -3 \\ 0 & 1 & 0 & | & 1 & -6 & -2 \\ 0 & 0 & 1 & | & 0 & -1 & 0 \end{bmatrix}.$$

Therefore, the required inverse is $\begin{bmatrix} 2 & -11 & -3 \\ 1 & -6 & -2 \\ 0 & -1 & 0 \end{bmatrix}.$

11.
$$\begin{bmatrix} 4 & 2 & 2 & | & 1 & 0 & 0 \\ -1 & -3 & 4 & | & 0 & 1 & 0 \\ 3 & -1 & 6 & | & 0 & 0 & 1 \end{bmatrix} \xrightarrow{R_1 - R_3} \begin{bmatrix} 1 & 3 & -4 & | & 1 & 0 & -1 \\ -1 & -3 & 4 & | & 0 & 1 & 0 \\ 3 & -1 & 6 & | & 0 & 0 & 1 \end{bmatrix}$$

$$\xrightarrow[R_3 + 5R_1]{R_2 + R_1} \begin{bmatrix} 1 & 3 & -4 & | & 1 & 0 & -1 \\ 0 & 0 & 0 & | & 1 & 1 & -1 \\ 3 & -1 & 6 & | & 0 & 0 & 1 \end{bmatrix}$$

Because there is a row of zeros to the left of the vertical line, we see that the inverse does not exist.

13.
$$\left[\begin{array}{ccc|ccc} 1 & 4 & -1 & 1 & 0 & 0 \\ 2 & 3 & -2 & 0 & 1 & 0 \\ -1 & 2 & 3 & 0 & 0 & 1 \end{array}\right] \xrightarrow[R_3+R_1]{R_2-2R_1} \left[\begin{array}{ccc|ccc} 1 & 4 & -1 & 1 & 0 & 0 \\ 0 & -5 & 0 & -2 & 1 & 0 \\ 0 & 6 & 2 & 1 & 0 & 1 \end{array}\right] \xrightarrow{R_2+R_3}$$

$$\left[\begin{array}{ccc|ccc} 1 & 4 & -1 & 1 & 0 & 0 \\ 0 & 1 & 2 & -1 & 1 & 1 \\ 0 & 6 & 2 & 1 & 0 & 1 \end{array}\right] \xrightarrow[R_3-6R_2]{R_1-4R_2} \left[\begin{array}{ccc|ccc} 1 & 0 & -9 & 5 & -4 & -4 \\ 0 & 1 & 2 & -1 & 1 & 1 \\ 0 & 0 & -10 & 7 & -6 & -5 \end{array}\right] \xrightarrow{-\frac{1}{10}R_3}$$

$$\left[\begin{array}{ccc|ccc} 1 & 0 & -9 & 5 & -4 & -4 \\ 0 & 1 & 2 & -1 & 1 & 1 \\ 0 & 0 & 1 & -\frac{7}{10} & \frac{3}{5} & \frac{1}{2} \end{array}\right] \xrightarrow[R_2-2R_3]{R_1+9R_3} \left[\begin{array}{ccc|ccc} 1 & 0 & 0 & -\frac{13}{10} & \frac{7}{5} & \frac{1}{2} \\ 0 & 1 & 0 & \frac{2}{5} & -\frac{1}{5} & 0 \\ 0 & 0 & 1 & -\frac{7}{10} & \frac{3}{5} & \frac{1}{2} \end{array}\right]$$

So $\quad A^{-1} = \left[\begin{array}{ccc} -\frac{13}{10} & \frac{7}{5} & \frac{1}{2} \\ \frac{2}{5} & -\frac{1}{5} & 0 \\ -\frac{7}{10} & \frac{3}{5} & \frac{1}{2} \end{array}\right].$

15.
$$\left[\begin{array}{cccc|cccc} 1 & 1 & -1 & 1 & 1 & 0 & 0 & 0 \\ 2 & 1 & 1 & 0 & 0 & 1 & 0 & 0 \\ 2 & 1 & 0 & 1 & 0 & 0 & 1 & 0 \\ 2 & -1 & -1 & 3 & 0 & 0 & 0 & 1 \end{array}\right] \xrightarrow[R_4-2R_1]{\substack{R_2-2R_1 \\ R_3-2R_1}} \left[\begin{array}{cccc|cccc} 1 & 1 & -1 & 1 & 1 & 0 & 0 & 0 \\ 0 & -1 & 3 & -2 & -2 & 1 & 0 & 0 \\ 0 & -1 & 2 & -1 & -2 & 0 & 1 & 0 \\ 0 & -3 & 1 & 1 & -2 & 0 & 0 & 1 \end{array}\right] \xrightarrow{-R_2}$$

$$\left[\begin{array}{cccc|cccc} 1 & 1 & -1 & 1 & 1 & 0 & 0 & 0 \\ 0 & 1 & -3 & 2 & 2 & -1 & 0 & 0 \\ 0 & -1 & 2 & -1 & -2 & 0 & 1 & 0 \\ 0 & -3 & 1 & 1 & -2 & 0 & 0 & 1 \end{array}\right] \xrightarrow[R_4+3R_2]{\substack{R_1-R_2 \\ R_3+R_2}} \left[\begin{array}{cccc|cccc} 1 & 0 & 2 & -1 & -1 & 1 & 0 & 0 \\ 0 & 1 & -3 & 2 & 2 & -1 & 0 & 0 \\ 0 & 0 & -1 & 1 & 0 & -1 & 1 & 0 \\ 0 & 0 & -8 & 7 & 4 & -3 & 0 & 1 \end{array}\right] \xrightarrow{-R_3}$$

$$\left[\begin{array}{cccc|cccc} 1 & 0 & 2 & -1 & -1 & 1 & 0 & 0 \\ 0 & 1 & -3 & 2 & 2 & -1 & 0 & 0 \\ 0 & 0 & 1 & -1 & 0 & 1 & -1 & 0 \\ 0 & 0 & -8 & 7 & 4 & -3 & 0 & 1 \end{array}\right] \xrightarrow[R_4+8R_3]{\substack{R_1-2R_3 \\ R_2+3R_3}} \left[\begin{array}{cccc|cccc} 1 & 0 & 0 & 1 & -1 & -1 & 2 & 0 \\ 0 & 1 & 0 & -1 & 2 & 2 & -3 & 0 \\ 0 & 0 & 1 & -1 & 0 & 1 & -1 & 0 \\ 0 & 0 & 0 & -1 & 4 & 5 & -8 & 1 \end{array}\right]$$

$$\begin{array}{c} R_1+R_4 \\ R_2-R_4 \\ R_3-R_4 \\ \xrightarrow{\;-R_4\;} \end{array} \left[\begin{array}{cccc|cccc} 1 & 0 & 0 & 0 & 3 & 4 & -6 & 1 \\ 0 & 1 & 0 & 0 & -2 & -3 & 5 & -1 \\ 0 & 0 & 1 & 0 & -4 & -4 & 7 & -1 \\ 0 & 0 & 0 & 1 & -4 & -5 & 8 & -1 \end{array}\right].$$

So the required inverse is

$$A^{-1} = \begin{bmatrix} 3 & 4 & -6 & 1 \\ -2 & -3 & 5 & -1 \\ -4 & -4 & 7 & -1 \\ -4 & -5 & 8 & -1 \end{bmatrix}.$$

We can verify our result by showing that $A^{-1}A = A$. Thus,

$$\begin{bmatrix} 3 & 4 & -6 & 1 \\ -2 & -3 & 5 & -1 \\ -4 & -4 & 7 & -1 \\ -4 & -5 & 8 & -1 \end{bmatrix} \begin{bmatrix} 1 & 1 & -1 & 1 \\ 2 & 1 & 1 & 0 \\ 2 & 1 & 0 & 1 \\ 2 & -1 & -1 & 3 \end{bmatrix} = \begin{bmatrix} 1 & 0 & 0 & 0 \\ 0 & 1 & 0 & 0 \\ 0 & 0 & 1 & 0 \\ 0 & 0 & 0 & 1 \end{bmatrix}.$$

17.   a. $A = \begin{bmatrix} 2 & 5 \\ 1 & 3 \end{bmatrix}$, $X = \begin{bmatrix} x \\ y \end{bmatrix}$, $B = \begin{bmatrix} 3 \\ 2 \end{bmatrix}$;

      b. $X = A^{-1}B = \begin{bmatrix} 3 & -5 \\ -1 & 2 \end{bmatrix}\begin{bmatrix} 3 \\ 2 \end{bmatrix} = \begin{bmatrix} -1 \\ 1 \end{bmatrix}$;

19.   a. $A = \begin{bmatrix} 2 & -3 & -4 \\ 0 & 0 & -1 \\ 1 & -2 & 1 \end{bmatrix}$, $X = \begin{bmatrix} x \\ y \\ z \end{bmatrix}$, $B = \begin{bmatrix} 4 \\ 3 \\ -8 \end{bmatrix}$

$$X = A^{-1}B = \begin{bmatrix} 2 & -11 & -3 \\ 1 & -6 & -2 \\ 0 & -1 & 0 \end{bmatrix}\begin{bmatrix} 4 \\ 3 \\ -8 \end{bmatrix} = \begin{bmatrix} -1 \\ 2 \\ -3 \end{bmatrix}$$

21. a. $A = \begin{bmatrix} 1 & 4 & -1 \\ 2 & 3 & -2 \\ -1 & 2 & 3 \end{bmatrix}$, $X = \begin{bmatrix} x \\ y \\ z \end{bmatrix}$, $B = \begin{bmatrix} 3 \\ 1 \\ 7 \end{bmatrix}$;

b. $X = A^{-1}B = \begin{bmatrix} -\frac{13}{10} & \frac{7}{5} & \frac{1}{2} \\ \frac{2}{5} & -\frac{1}{5} & 0 \\ -\frac{7}{10} & \frac{3}{5} & \frac{1}{2} \end{bmatrix} \begin{bmatrix} 3 \\ 1 \\ 7 \end{bmatrix} = \begin{bmatrix} 1 \\ 1 \\ 2 \end{bmatrix}$.

23. a. $A = \begin{bmatrix} 1 & 1 & -1 & 1 \\ 2 & 1 & 1 & 0 \\ 2 & 1 & 0 & 1 \\ 2 & -1 & -1 & 3 \end{bmatrix}$, $X = \begin{bmatrix} x_1 \\ x_2 \\ x_3 \\ x_4 \end{bmatrix}$, $B = \begin{bmatrix} 6 \\ 4 \\ 7 \\ 9 \end{bmatrix}$.

b. $X = A^{-1}B = \begin{bmatrix} 3 & 4 & -6 & 1 \\ -2 & -3 & 5 & 1 \\ -4 & -4 & 7 & -1 \\ -4 & -5 & 8 & -1 \end{bmatrix} \begin{bmatrix} 6 \\ 4 \\ 7 \\ 9 \end{bmatrix} = \begin{bmatrix} 1 \\ 2 \\ 0 \\ 3 \end{bmatrix}$.

25. a-b. $A = \begin{bmatrix} 1 & 2 \\ 2 & -1 \end{bmatrix}$, $X = \begin{bmatrix} x \\ y \end{bmatrix}$, $B = \begin{bmatrix} b_1 \\ b_2 \end{bmatrix}$;

(i). $X = A^{-1}B = \begin{bmatrix} 0.2 & 0.4 \\ 0.4 & -0.2 \end{bmatrix} \begin{bmatrix} 14 \\ 5 \end{bmatrix} = \begin{bmatrix} 4.8 \\ 4.6 \end{bmatrix}$ and we conclude that $x = 4.8$ and $y = 4.6$.

(ii). $X = A^{-1}B = \begin{bmatrix} 0.2 & 0.4 \\ 0.4 & -0.2 \end{bmatrix} \begin{bmatrix} 4 \\ -1 \end{bmatrix} = \begin{bmatrix} 0.4 \\ 1.8 \end{bmatrix}$ and we conclude that $x = 0.4$ and $y = 1.8$.

27. a. First we find A^{-1}.

$$\begin{bmatrix} 1 & 2 & 1 & | & 1 & 0 & 0 \\ 1 & 1 & 1 & | & 0 & 1 & 0 \\ 3 & 1 & 1 & | & 0 & 0 & 1 \end{bmatrix} \xrightarrow[R_3-3R_1]{R_2-R_1} \begin{bmatrix} 1 & 2 & 1 & | & 1 & 0 & 0 \\ 0 & -1 & 0 & | & -1 & 1 & 0 \\ 0 & -5 & -2 & | & -3 & 0 & 1 \end{bmatrix} \xrightarrow{-R_2}$$

$$\begin{bmatrix} 1 & 2 & 1 & 1 & 0 & 0 \\ 0 & 1 & 0 & 1 & -1 & 0 \\ 0 & -5 & -2 & -3 & 0 & 1 \end{bmatrix} \xrightarrow[R_3+5R_2]{R_1-2R_2} \begin{bmatrix} 1 & 0 & 1 & -1 & 2 & 0 \\ 0 & 1 & 0 & 1 & -1 & 0 \\ 0 & 0 & -2 & 2 & -5 & 1 \end{bmatrix} \xrightarrow{-\frac{1}{2}R_3}$$

$$\begin{bmatrix} 1 & 0 & 1 & -1 & 2 & 0 \\ 0 & 1 & 0 & 1 & -1 & 0 \\ 0 & 0 & 1 & -1 & \frac{5}{2} & -\frac{1}{2} \end{bmatrix} \xrightarrow{R_1-R_3} \begin{bmatrix} 1 & 0 & 0 & 0 & -\frac{1}{2} & \frac{1}{2} \\ 0 & 1 & 0 & 1 & -1 & 0 \\ 0 & 0 & 1 & -1 & \frac{5}{2} & -\frac{1}{2} \end{bmatrix}$$

$$\begin{bmatrix} 1 & 2 & 1 \\ 1 & 1 & 1 \\ 3 & 1 & 1 \end{bmatrix} \begin{bmatrix} x \\ y \\ z \end{bmatrix} = \begin{bmatrix} b_1 \\ b_2 \\ b_3 \end{bmatrix}$$

(i). $\begin{bmatrix} x \\ y \\ z \end{bmatrix} = \begin{bmatrix} 0 & -\frac{1}{2} & \frac{1}{2} \\ 1 & -1 & 0 \\ -1 & \frac{5}{2} & -\frac{1}{2} \end{bmatrix} \begin{bmatrix} 7 \\ 4 \\ 2 \end{bmatrix} = \begin{bmatrix} -1 \\ 3 \\ 2 \end{bmatrix}$

and we conclude that $x = -1$, $y = 3$, and $z = 2$.

(ii). $\begin{bmatrix} x \\ y \\ z \end{bmatrix} = \begin{bmatrix} 0 & -\frac{1}{2} & \frac{1}{2} \\ 1 & -1 & 0 \\ -1 & \frac{5}{2} & -\frac{1}{2} \end{bmatrix} \begin{bmatrix} 5 \\ -3 \\ -1 \end{bmatrix} = \begin{bmatrix} 1 \\ 8 \\ -12 \end{bmatrix}$

and we conclude that $x = 1$, $y = 8$, and $z = -12$.

29. a. $\begin{bmatrix} 3 & 2 & -1 & 1 & 0 & 0 \\ 2 & -3 & 1 & 0 & 1 & 0 \\ 1 & -1 & -1 & 0 & 0 & 1 \end{bmatrix} \xrightarrow{R_1 \leftrightarrow R_3} \begin{bmatrix} 1 & -1 & -1 & 0 & 0 & 1 \\ 2 & -3 & 1 & 0 & 1 & 0 \\ 3 & 2 & -1 & 1 & 0 & 0 \end{bmatrix} \xrightarrow[R_3-3R_1]{R_2-2R_1}$

$$\begin{bmatrix} 1 & -1 & -1 & 0 & 0 & 1 \\ 0 & -1 & 3 & 0 & 1 & -2 \\ 0 & 5 & 2 & 1 & 0 & -3 \end{bmatrix} \xrightarrow{-R_2} \begin{bmatrix} 1 & -1 & -1 & 0 & 0 & 1 \\ 0 & 1 & -3 & 0 & -1 & 2 \\ 0 & 5 & 2 & 1 & 0 & -3 \end{bmatrix} \xrightarrow[R_3-5R_2]{R_1+R_2}$$

$$\begin{bmatrix} 1 & 0 & -4 & 0 & -1 & 3 \\ 0 & 1 & -3 & 0 & -1 & 2 \\ 0 & 0 & 17 & 1 & 5 & -13 \end{bmatrix} \xrightarrow{\frac{1}{17}R_3} \begin{bmatrix} 1 & 0 & -4 & 0 & -1 & 3 \\ 0 & 1 & -3 & 0 & -1 & 2 \\ 0 & 0 & 1 & \frac{1}{17} & \frac{5}{17} & -\frac{13}{17} \end{bmatrix}$$

$$\begin{array}{c} R_1+4R_3 \\ \xrightarrow{R_2+3R_3} \end{array} \begin{bmatrix} 1 & 0 & 0 & | & \frac{4}{17} & \frac{3}{17} & -\frac{1}{17} \\ 0 & 1 & 0 & | & \frac{3}{17} & -\frac{2}{17} & -\frac{5}{17} \\ 0 & 0 & 1 & | & \frac{1}{17} & \frac{5}{17} & -\frac{13}{17} \end{bmatrix}.$$

Therefore $A^{-1} = \begin{bmatrix} \frac{4}{17} & \frac{3}{17} & -\frac{1}{17} \\ \frac{3}{17} & -\frac{2}{17} & -\frac{5}{17} \\ \frac{1}{17} & \frac{5}{17} & -\frac{13}{17} \end{bmatrix}.$

Next, $\begin{bmatrix} 3 & 2 & -1 \\ 2 & -3 & 1 \\ 1 & -1 & -1 \end{bmatrix} \begin{bmatrix} x \\ y \\ z \end{bmatrix} = \begin{bmatrix} b_1 \\ b_2 \\ b_3 \end{bmatrix}$

(i). $\begin{bmatrix} x \\ y \\ z \end{bmatrix} = \begin{bmatrix} \frac{4}{17} & \frac{3}{17} & -\frac{1}{17} \\ \frac{3}{17} & -\frac{2}{17} & -\frac{5}{17} \\ \frac{1}{17} & \frac{5}{17} & -\frac{13}{17} \end{bmatrix} \begin{bmatrix} 2 \\ -2 \\ 4 \end{bmatrix} = \begin{bmatrix} -\frac{2}{17} \\ -\frac{10}{17} \\ -\frac{60}{17} \end{bmatrix}$

We conclude that $x = -2/17$, $y = -10/17$, and $z = -60/17$.

(ii). $\begin{bmatrix} x \\ y \\ z \end{bmatrix} = \begin{bmatrix} \frac{4}{17} & \frac{3}{17} & -\frac{1}{17} \\ \frac{3}{17} & -\frac{2}{17} & -\frac{5}{17} \\ \frac{1}{17} & \frac{5}{17} & -\frac{13}{17} \end{bmatrix} \begin{bmatrix} 8 \\ -3 \\ 6 \end{bmatrix} = \begin{bmatrix} 1 \\ 0 \\ -5 \end{bmatrix}$

We conclude that $x = 1$, $y = 0$, and $z = -5$.

31.   a. $AX = B_1$ and $AX = B_2$, where

$$A = \begin{bmatrix} 1 & 1 & 1 & 1 \\ 1 & -1 & -1 & 1 \\ 0 & 1 & 2 & 2 \\ 1 & 2 & 1 & -2 \end{bmatrix}, \ X = \begin{bmatrix} x_1 \\ x_2 \\ x_3 \\ x_4 \end{bmatrix}, \ B_1 = \begin{bmatrix} 1 \\ -1 \\ 4 \\ 0 \end{bmatrix} \ \text{and} \ B_2 = \begin{bmatrix} 2 \\ 8 \\ 4 \\ -1 \end{bmatrix}.$$

We first find A^{-1}.

$$\begin{bmatrix} 1 & 1 & 1 & 1 & | & 1 & 0 & 0 & 0 \\ 1 & -1 & -1 & 1 & | & 0 & 1 & 0 & 0 \\ 0 & 1 & 2 & 2 & | & 0 & 0 & 1 & 0 \\ 1 & 2 & 1 & -2 & | & 0 & 0 & 0 & 1 \end{bmatrix} \xrightarrow[R_4-R_1]{R_2-R_1} \begin{bmatrix} 1 & 1 & 1 & 1 & | & 1 & 0 & 0 & 0 \\ 0 & -2 & -2 & 0 & | & -1 & 1 & 0 & 0 \\ 0 & 1 & 2 & 2 & | & 0 & 0 & 1 & 0 \\ 0 & 1 & 0 & -3 & | & -1 & 0 & 0 & 1 \end{bmatrix}$$

$$\xrightarrow{R_2 \leftrightarrow R_3} \begin{bmatrix} 1 & 1 & 1 & 1 & | & 1 & 0 & 0 & 0 \\ 0 & 1 & 2 & 2 & | & 0 & 0 & 1 & 0 \\ 0 & -2 & -2 & 0 & | & -1 & 1 & 0 & 0 \\ 0 & 1 & 0 & -3 & | & -1 & 0 & 0 & 1 \end{bmatrix} \xrightarrow[\substack{R_3+2R_2 \\ R_4-R_2}]{R_1-R_2}$$

$$\begin{bmatrix} 1 & 0 & -1 & -1 & | & 1 & 0 & -1 & 0 \\ 0 & 1 & 2 & 2 & | & 0 & 0 & 1 & 0 \\ 0 & 0 & 2 & 4 & | & -1 & 1 & 2 & 0 \\ 0 & 0 & -2 & -5 & | & -1 & 0 & -1 & 1 \end{bmatrix} \xrightarrow{\frac{1}{2}R_3} \begin{bmatrix} 1 & 0 & -1 & -1 & | & 1 & 0 & -1 & 0 \\ 0 & 1 & 2 & 2 & | & 0 & 0 & 1 & 0 \\ 0 & 0 & 1 & 2 & | & -\frac{1}{2} & \frac{1}{2} & 1 & 0 \\ 0 & 0 & -2 & -5 & | & -1 & 0 & -1 & 1 \end{bmatrix}$$

$$\xrightarrow[\substack{R_2-2R_3 \\ R_4+2R_3}]{R_1+R_3} \begin{bmatrix} 1 & 0 & 0 & 1 & | & \frac{1}{2} & \frac{1}{2} & 0 & 0 \\ 0 & 1 & 0 & -2 & | & 1 & -1 & -1 & 0 \\ 0 & 0 & 1 & 2 & | & -\frac{1}{2} & \frac{1}{2} & 1 & 0 \\ 0 & 0 & 0 & -1 & | & -2 & 1 & 1 & 1 \end{bmatrix} \xrightarrow[\substack{R_2-2R_4 \\ R_3+2R_4 \\ -R_4}]{R_1+R_4}$$

$$\begin{bmatrix} 1 & 0 & 0 & 0 & | & -\frac{3}{2} & \frac{3}{2} & 1 & 1 \\ 0 & 1 & 0 & 0 & | & 5 & -3 & -3 & -2 \\ 0 & 0 & 1 & 0 & | & -\frac{9}{2} & \frac{5}{2} & 3 & 2 \\ 0 & 0 & 0 & 1 & | & 2 & -1 & -1 & -1 \end{bmatrix}. \quad \text{So} \quad A^{-1} = \begin{bmatrix} -\frac{3}{2} & \frac{3}{2} & 1 & 1 \\ 5 & -3 & -3 & -2 \\ -\frac{9}{2} & \frac{5}{2} & 3 & 2 \\ 2 & -1 & -1 & -1 \end{bmatrix}.$$

b. (i). $\begin{bmatrix} x_1 \\ x_2 \\ x_3 \\ x_4 \end{bmatrix} = \begin{bmatrix} -\frac{3}{2} & \frac{3}{2} & 1 & 1 \\ 5 & -3 & -3 & -2 \\ -\frac{9}{2} & \frac{5}{2} & 3 & 2 \\ 2 & -1 & -1 & -1 \end{bmatrix} \begin{bmatrix} 1 \\ -1 \\ 4 \\ 0 \end{bmatrix} = \begin{bmatrix} 1 \\ -4 \\ 5 \\ -1 \end{bmatrix}$

and we conclude that $x_1 = 1$, $x_2 = -4$, $x_3 = 5$, and $x_4 = -1$.

(ii).
$$\begin{bmatrix} x_1 \\ x_2 \\ x_3 \\ x_4 \end{bmatrix} = \begin{bmatrix} -\frac{3}{2} & \frac{3}{2} & 1 & 1 \\ 5 & -3 & -3 & -2 \\ -\frac{9}{2} & \frac{5}{2} & 3 & 2 \\ 2 & -1 & -1 & -1 \end{bmatrix} \begin{bmatrix} 2 \\ 8 \\ 4 \\ -1 \end{bmatrix} = \begin{bmatrix} 12 \\ -24 \\ 21 \\ -7 \end{bmatrix}$$

and we conclude that $x_1 = 12$, $x_2 = -24$, $x_3 = 21$, and $x_4 = -7$.

33. a. Using Formula (13), we find

$$A^{-1} = \frac{1}{(2)(-5)-(-4)(3)} \begin{bmatrix} -5 & -3 \\ 4 & 2 \end{bmatrix} = \begin{bmatrix} -\frac{5}{2} & -\frac{3}{2} \\ 2 & 1 \end{bmatrix}.$$

b. Using Formula (13) once again, we find

$$\left(A^{-1}\right)^{-1} = \frac{1}{\left(-\frac{5}{2}\right)(1)-2\left(-\frac{3}{2}\right)} \begin{bmatrix} 1 & \frac{3}{2} \\ -2 & -\frac{5}{2} \end{bmatrix} = \begin{bmatrix} 2 & 3 \\ -4 & -5 \end{bmatrix} = A.$$

35. a. $ABC = \begin{bmatrix} 2 & -5 \\ 1 & -3 \end{bmatrix} \begin{bmatrix} 4 & 3 \\ 1 & 1 \end{bmatrix} \begin{bmatrix} 2 & 3 \\ -2 & 1 \end{bmatrix}$

$$= \begin{bmatrix} 2 & -5 \\ 1 & -3 \end{bmatrix} \begin{bmatrix} 2 & 15 \\ 0 & 4 \end{bmatrix} = \begin{bmatrix} 4 & 10 \\ 2 & 3 \end{bmatrix}.$$

Using the formula for finding the inverse of a 2×2 matrix, we find

$$A^{-1} = \begin{bmatrix} 3 & -5 \\ 1 & -2 \end{bmatrix}, \quad B^{-1} = \begin{bmatrix} 1 & -3 \\ -1 & 4 \end{bmatrix}, \quad C^{-1} = \begin{bmatrix} \frac{1}{8} & -\frac{3}{8} \\ \frac{1}{4} & \frac{1}{4} \end{bmatrix}.$$

b. Using the formula for finding the inverse of a 2×2 matrix, we find

$$(ABC)^{-1} = \begin{bmatrix} -\frac{3}{8} & \frac{5}{4} \\ \frac{1}{4} & -\frac{1}{2} \end{bmatrix}$$

$$C^{-1}B^{-1}A^{-1} = \begin{bmatrix} \frac{1}{8} & -\frac{3}{8} \\ \frac{1}{4} & \frac{1}{4} \end{bmatrix} \begin{bmatrix} 1 & -3 \\ -1 & 4 \end{bmatrix} \begin{bmatrix} 3 & -5 \\ 1 & -2 \end{bmatrix}$$

2 Systems of Linear Equations and Matrices

$$= \begin{bmatrix} \frac{1}{8} & -\frac{3}{8} \\ \frac{1}{4} & \frac{1}{4} \end{bmatrix} \begin{bmatrix} 0 & 1 \\ 1 & -3 \end{bmatrix} = \begin{bmatrix} -\frac{3}{8} & \frac{5}{4} \\ \frac{1}{4} & -\frac{1}{2} \end{bmatrix}.$$

Therefore, $(ABC)^{-1} = C^{-1}B^{-1}A^{-1}$.

37. Let x denote the number of copies of the deluxe edition and y the number of copies of the standard edition demanded per month when the unit prices are p and q dollars, respectively. Then the three systems of linear equations

$$\begin{array}{lll} 5x + y = 20000 & 5x + y = 25000 & 5x + y = 25000 \\ x + 3y = 15000 & x + 3y = 15000 & x + 3y = 20000 \end{array}$$

give the quantity demanded of each edition at the stated price. These systems may be written in the form $AX = B_1$, $AX = B_2$, and $AX = B_3$, where

$$A = \begin{bmatrix} 5 & 1 \\ 1 & 3 \end{bmatrix}, \quad B_1 = \begin{bmatrix} 20000 \\ 15000 \end{bmatrix}, \quad B_2 = \begin{bmatrix} 25000 \\ 15000 \end{bmatrix}, \quad \text{and} \quad B_3 = \begin{bmatrix} 25000 \\ 20000 \end{bmatrix}.$$

Using the formula for finding the inverse of a 2 × 2 matrix, with $a = 5$, $b = 1$, $c = 1$, $d = 3$, and $D = ad - bc = (5)(3) - (1)(1) = 14$, we find that

$$A^{-1} = \begin{bmatrix} \frac{3}{14} & -\frac{1}{14} \\ -\frac{1}{14} & \frac{5}{14} \end{bmatrix}.$$

a. $\begin{bmatrix} x \\ y \end{bmatrix} = \begin{bmatrix} \frac{3}{14} & -\frac{1}{14} \\ -\frac{1}{14} & \frac{5}{14} \end{bmatrix} \begin{bmatrix} 20,000 \\ 15,000 \end{bmatrix} = \begin{bmatrix} 3,214 \\ 3,929 \end{bmatrix}$ b. $\begin{bmatrix} x \\ y \end{bmatrix} = \begin{bmatrix} \frac{3}{14} & -\frac{1}{14} \\ -\frac{1}{14} & \frac{5}{14} \end{bmatrix} \begin{bmatrix} 25,000 \\ 15,000 \end{bmatrix} = \begin{bmatrix} 4,286 \\ 3,571 \end{bmatrix}$

c. $\begin{bmatrix} x \\ y \end{bmatrix} = \begin{bmatrix} \frac{3}{14} & -\frac{1}{14} \\ -\frac{1}{14} & \frac{5}{14} \end{bmatrix} \begin{bmatrix} 25,000 \\ 20,000 \end{bmatrix} = \begin{bmatrix} 3,929 \\ 5,357 \end{bmatrix}.$

39. Let x, y, and z (in millions of dollars) be the amount awarded to organization I, II, and III, respectively. Then we have

$$\begin{array}{ll} 0.6x + 0.4y + 0.2z = 9.2 & (8.2) \\ 0.3x + 0.3y + 0.6z = 9.6 & (7.2) \\ 0.1x + 0.3y + 0.2z = 5.2 & (3.6). \end{array}$$

The quantities within the brackets are for part (b). We can rewrite the systems as $AX = B_1$, and $AX = B_2$. Put

$$X = \begin{bmatrix} x \\ y \\ z \end{bmatrix}, \quad A = \begin{bmatrix} 6 & 4 & 2 \\ 3 & 3 & 6 \\ 1 & 3 & 2 \end{bmatrix}, \quad B_1 = \begin{bmatrix} 92 \\ 96 \\ 52 \end{bmatrix}, \quad \text{and} \quad B_2 = \begin{bmatrix} 82 \\ 72 \\ 36 \end{bmatrix}.$$

To find A^{-1}, we use the Gauss-Jordan method:

$$\begin{bmatrix} 6 & 4 & 2 & | & 1 & 0 & 0 \\ 3 & 3 & 6 & | & 0 & 1 & 0 \\ 1 & 3 & 2 & | & 0 & 0 & 1 \end{bmatrix} \xrightarrow{R_1 \leftrightarrow R_3} \begin{bmatrix} 1 & 3 & 2 & | & 0 & 0 & 1 \\ 3 & 3 & 6 & | & 0 & 1 & 0 \\ 6 & 4 & 2 & | & 1 & 0 & 0 \end{bmatrix} \xrightarrow[R_3 - 6R_1]{R_2 - 3R_1}$$

$$\begin{bmatrix} 1 & 3 & 2 & | & 0 & 0 & 1 \\ 0 & -6 & 0 & | & 0 & 1 & -3 \\ 0 & -14 & -10 & | & 1 & 0 & -6 \end{bmatrix} \xrightarrow{-\frac{1}{6}R_2} \begin{bmatrix} 1 & 3 & 2 & | & 0 & 0 & 1 \\ 0 & 1 & 0 & | & 0 & -\frac{1}{6} & \frac{1}{2} \\ 0 & -14 & -10 & | & 1 & 0 & -6 \end{bmatrix} \xrightarrow[R_3 + 14R_2]{R_1 - 3R_2}$$

$$\begin{bmatrix} 1 & 0 & 2 & | & 0 & \frac{1}{2} & -\frac{1}{2} \\ 0 & 1 & 0 & | & 0 & -\frac{1}{6} & \frac{1}{2} \\ 0 & 0 & -10 & | & 1 & -\frac{7}{3} & 1 \end{bmatrix} \xrightarrow{-\frac{1}{10}R_3} \begin{bmatrix} 1 & 0 & 2 & | & 1 & \frac{1}{2} & -\frac{1}{2} \\ 0 & 1 & 0 & | & 0 & -\frac{1}{6} & \frac{1}{2} \\ 0 & 0 & 1 & | & -\frac{1}{10} & \frac{7}{30} & -\frac{1}{10} \end{bmatrix} \xrightarrow{R_1 - 2R_3}$$

$$\begin{bmatrix} 1 & 0 & 0 & | & \frac{1}{5} & \frac{1}{30} & -\frac{3}{10} \\ 0 & 1 & 0 & | & 0 & -\frac{1}{6} & \frac{1}{2} \\ 0 & 0 & 1 & | & -\frac{1}{10} & \frac{7}{30} & -\frac{1}{10} \end{bmatrix}.$$

a. $$X = A^{-1}B_1 = \begin{bmatrix} \frac{1}{5} & \frac{1}{30} & -\frac{3}{10} \\ 0 & -\frac{1}{6} & \frac{1}{2} \\ -\frac{1}{10} & \frac{7}{30} & -\frac{1}{10} \end{bmatrix} \begin{bmatrix} 92 \\ 96 \\ 52 \end{bmatrix} = \begin{bmatrix} 6 \\ 10 \\ 8 \end{bmatrix}$$

that is, $x = 6$, $y = 10$, and $z = 8$, and Organization I will receive $6 million, Organization II will receive $10 million, and Organization III will receive $8 million.

b. $$X = A^{-1}B_1 = \begin{bmatrix} \frac{1}{5} & \frac{1}{30} & -\frac{3}{10} \\ 0 & -\frac{1}{6} & \frac{1}{2} \\ -\frac{1}{10} & \frac{7}{30} & -\frac{1}{10} \end{bmatrix} \begin{bmatrix} 82 \\ 72 \\ 36 \end{bmatrix} = \begin{bmatrix} 8 \\ 6 \\ 5 \end{bmatrix}$$

that is, $x = 8$, $y = 6$, and $z = 5$, and Organization I will receive \$8 million, Organization II will receive \$6 million, and Organization III will receive \$5 million.

41. False. The matrix A has a multiplicative inverse if and only if $ad - bc \neq 0$, in which case case

$$A^{-1} = \frac{1}{ad - bc} \begin{bmatrix} d & -b \\ -c & a \end{bmatrix}.$$

43. **Case 1:** $a \neq 0$

$$\begin{bmatrix} a & b & | & 1 & 0 \\ c & d & | & 0 & 1 \end{bmatrix} \xrightarrow{\frac{1}{a}R_1} \begin{bmatrix} a & \frac{b}{a} & | & \frac{1}{a} & 0 \\ c & d & | & 0 & 1 \end{bmatrix} \xrightarrow{R_2 - cR_1} \begin{bmatrix} 1 & \frac{b}{a} & | & \frac{1}{a} & 0 \\ 0 & d - \frac{bc}{a} & | & -\frac{c}{a} & 1 \end{bmatrix}$$

$$\xrightarrow{\frac{a}{ad-bc}R_2} \begin{bmatrix} 1 & \frac{b}{a} & | & \frac{1}{a} & 0 \\ 0 & 1 & | & -\frac{c}{ad-bc} & \frac{a}{ad-bc} \end{bmatrix} \xrightarrow{R_1 - \frac{b}{a}R_2} \begin{bmatrix} 1 & 0 & | & \frac{d}{ad-bc} & -\frac{b}{ad-bc} \\ 0 & 1 & | & -\frac{c}{ad-bc} & \frac{a}{ad-bc} \end{bmatrix}$$

since $\dfrac{1}{a} - \dfrac{b}{a}\left(-\dfrac{c}{ad - bc}\right) = \dfrac{ad - bc + bc}{a(ad - bc)} = \dfrac{d}{ad - bc}$ provided $ad - bc \neq 0$.

Case 2: $a = 0$

$$\begin{bmatrix} 0 & b & | & 1 & 0 \\ c & d & | & 0 & 1 \end{bmatrix} \xrightarrow{R_1 \leftrightarrow R_2} \begin{bmatrix} c & d & | & 0 & 1 \\ 0 & b & | & 1 & 0 \end{bmatrix} \xrightarrow[\frac{1}{b}R_2]{\frac{1}{c}R_1}$$

$$\begin{bmatrix} 1 & \frac{d}{c} & | & 1 & \frac{1}{c} \\ 0 & 1 & | & \frac{1}{b} & 0 \end{bmatrix} \xrightarrow{R_1 - \frac{d}{c}R_2} \begin{bmatrix} 1 & 0 & | & -\frac{d}{bc} & \frac{1}{c} \\ 0 & 1 & | & \frac{1}{b} & 0 \end{bmatrix}$$ provided $bc \neq 0$.

USING TECHNOLOGY EXERCISES 2.6, page 161

1.
$$\begin{bmatrix} 0.36 & 0.04 & -0.36 \\ 0.06 & 0.05 & 0.20 \\ -0.19 & 0.10 & 0.09 \end{bmatrix}$$

3.
$$\begin{bmatrix} 0.01 & -0.09 & 0.31 & -0.11 \\ -0.25 & 0.58 & -0.15 & -0.02 \\ 0.86 & -0.42 & 0.07 & -0.37 \\ -0.27 & 0.01 & -0.05 & 0.31 \end{bmatrix}$$

$$\begin{bmatrix} 0.30 & 0.85 & -0.10 & -0.77 & -0.11 \\ -0.21 & 0.10 & 0.01 & -0.26 & 0.21 \\ 0.03 & -0.16 & 0.12 & -0.01 & 0.03 \\ -0.14 & -0.46 & 0.13 & 0.71 & -0.05 \\ 0.10 & -0.05 & -0.10 & -0.03 & 0.11 \end{bmatrix}$$

5.

EXERCISES 2.7, page 171

1.   a. The amount of agricultural products consumed in the production of $100 million worth of manufactured goods is given by (100)(0.10), or $10 million.

b. The amount of manufactured goods required to produce $200 million of all goods in the economy is given by $200(0.1 + 0.4 + 0.3) = 160$, or $160 million.

c. From the input-output matrix, we see that the agricultural sector consumes the greatest amount of agricultural products, namely, 0.4 units, in the production of each unit of goods in that sector. The manufacturing and transportation sectors consume the least, 0.1 units each.

3.   Multiplying both sides of the given equation on the left by $(I - A)^{-1}$, we see that
$$X = (I - A)^{-1}D.$$

Now, $(I - A) = \begin{bmatrix} 1 & 0 \\ 0 & 1 \end{bmatrix} - \begin{bmatrix} 0.4 & 0.2 \\ 0.3 & 0.1 \end{bmatrix} = \begin{bmatrix} 0.6 & -0.2 \\ -0.3 & 0.9 \end{bmatrix}.$

Using the formula for finding the inverse of a 2×2 matrix, we find
$$(I - A)^{-1} = \begin{bmatrix} 1.875 & 0.417 \\ 0.625 & 1.25 \end{bmatrix}.$$

Then, $(I - A)^{-1}X = \begin{bmatrix} 1.875 & 0.417 \\ 0.625 & 1.25 \end{bmatrix}\begin{bmatrix} 10 \\ 12 \end{bmatrix} = \begin{bmatrix} 23.754 \\ 21.25 \end{bmatrix}.$

5.   We first compute

$$(I - A) = \begin{bmatrix} 1 & 0 \\ 0 & 1 \end{bmatrix} - \begin{bmatrix} 0.5 & 0.2 \\ 0.2 & 0.5 \end{bmatrix} = \begin{bmatrix} 0.5 & -0.2 \\ -0.2 & 0.5 \end{bmatrix}$$

Using the formula for finding the inverse of a 2×2 matrix, we find

$$(I - A)^{-1} = \begin{bmatrix} 2.381 & 0.952 \\ 0.952 & 2.381 \end{bmatrix}.$$

Then $\begin{bmatrix} x \\ y \end{bmatrix} = \begin{bmatrix} 2.381 & 0.952 \\ 0.952 & 2.381 \end{bmatrix} \begin{bmatrix} 10 \\ 20 \end{bmatrix} = \begin{bmatrix} 42.85 \\ 57.14 \end{bmatrix}$

7.   We verify

$$(I - A)(I - A)^{-1} = \begin{bmatrix} 0.92 & -0.60 & -0.30 \\ -0.04 & 0.98 & -0.01 \\ -0.02 & 0 & 0.94 \end{bmatrix} \begin{bmatrix} 1.13 & 0.69 & 0.37 \\ 0.05 & 1.05 & 0.03 \\ 0.02 & 0.02 & 1.07 \end{bmatrix} = \begin{bmatrix} 1 & 0 & 0 \\ 0 & 1 & 0 \\ 0 & 0 & 1 \end{bmatrix}$$

9.   a. $A = \begin{bmatrix} 0.2 & 0.4 \\ 0.3 & 0.3 \end{bmatrix}$ and

$$(I - A) = \begin{bmatrix} 1 & 0 \\ 0 & 1 \end{bmatrix} - \begin{bmatrix} 0.2 & 0.4 \\ 0.3 & 0.3 \end{bmatrix} = \begin{bmatrix} 0.8 & -0.4 \\ -0.3 & 0.7 \end{bmatrix}.$$

Using the formula for finding the inverse of a 2×2 matrix, we find

$$(I - A)^{-1} = \begin{bmatrix} 1.591 & 0.909 \\ 0.682 & 1.818 \end{bmatrix}.$$

Then $\begin{bmatrix} x \\ y \end{bmatrix} = \begin{bmatrix} 1.591 & 0.909 \\ 0.682 & 1.818 \end{bmatrix} \begin{bmatrix} 120 \\ 140 \end{bmatrix} = \begin{bmatrix} 318.18 \\ 336.36 \end{bmatrix}$

To fullfill consumer demand, $318.2 million worth of agricultural goods and $336.4 million worth of manufactured goods should be produced.

b. The net value of goods consumed in the internal process of production is

$$AX = X - D = \begin{bmatrix} 318.18 \\ 336.36 \end{bmatrix} - \begin{bmatrix} 120 \\ 140 \end{bmatrix} = \begin{bmatrix} 198.18 \\ 196.36 \end{bmatrix}.$$

or $198.2 million of agricultural goods and $196.4 million worth of manufactured goods.

11. a.

$$(I - A) = \begin{bmatrix} 1 & 0 & 0 \\ 0 & 1 & 0 \\ 0 & 0 & 1 \end{bmatrix} - \begin{bmatrix} 0.4 & 0.1 & 0.1 \\ 0.1 & 0.4 & 0.3 \\ 0.2 & 0.2 & 0.2 \end{bmatrix} = \begin{bmatrix} 0.6 & -0.1 & -0.1 \\ -0.1 & 0.6 & -0.3 \\ -0.2 & -0.2 & 0.8 \end{bmatrix}$$

Using the methods of Section 2.6 we next compute the inverse of $(1 - A)^{-1}$ and use this value to find

$$X = (1 - A)^{-1} D = \begin{bmatrix} 1.875 & 0.446 & 0.402 \\ 0.625 & 2.054 & 0.848 \\ 0.625 & 0.625 & 1.563 \end{bmatrix} \begin{bmatrix} 200 \\ 100 \\ 60 \end{bmatrix} = \begin{bmatrix} 443.7 \\ 381.3 \\ 281.3 \end{bmatrix}.$$

Therefore, to fulfull demand, $443.7 million worth of agricultural products, $381.3 million worth of manufactured products, and $281.3 million worth of transportation services should be produced.

b. To meet the gross output, the value of goods and transportation consumed in the internal process of production is

$$AX = X - D = \begin{bmatrix} 443.7 \\ 381.3 \\ 281.3 \end{bmatrix} - \begin{bmatrix} 200 \\ 100 \\ 60 \end{bmatrix} = \begin{bmatrix} 243.7 \\ 281.3 \\ 221.3 \end{bmatrix},$$

or $243.7 million worth of agricultural products, $281.3 million worth of manufactured services, and $221.3 million worth of transportation services.

13. We want to solve the equation $(I - A)X = D$ for X, the total output matrix. First, we compute

$$(I - A) = \begin{bmatrix} 1 & 0 \\ 0 & 1 \end{bmatrix} - \begin{bmatrix} 0.4 & 0.2 \\ 0.3 & 0.5 \end{bmatrix} = \begin{bmatrix} 0.6 & -0.2 \\ -0.3 & 0.5 \end{bmatrix}.$$

Using the formula for finding the inverse of a 2×2 matrix, we find

$$(I - A)^{-1} = \begin{bmatrix} 2.08 & 0.833 \\ 1.25 & 2.5 \end{bmatrix}$$

Therefore,

$$X = (I - A)^{-1} D = \begin{bmatrix} 2.08 & 0.833 \\ 1.25 & 2.5 \end{bmatrix} \begin{bmatrix} 12 \\ 24 \end{bmatrix} = \begin{bmatrix} 45 \\ 75 \end{bmatrix}.$$

We conclude that $45 million worth of goods of one industry and $75 million worth of goods of the other industry must be produced.

15. First, we compute

$$I - A = \begin{bmatrix} 1 & 0 & 0 \\ 0 & 1 & 0 \\ 0 & 0 & 1 \end{bmatrix} - \begin{bmatrix} 0.2 & 0.4 & 0.2 \\ 0.5 & 0 & 0.5 \\ 0 & 0.2 & 0 \end{bmatrix} = \begin{bmatrix} 0.8 & -0.4 & -0.2 \\ -0.5 & 1 & -0.5 \\ 0 & -0.2 & 1 \end{bmatrix}.$$

Next, using the Gauss-Jordan method, we find

$$(I - A)^{-1} = \begin{bmatrix} 1.8 & 0.88 & 0.80 \\ 1 & 1.6 & 1 \\ 0.2 & 0.32 & 1.20 \end{bmatrix}$$

Then

$$\begin{bmatrix} x \\ y \\ z \end{bmatrix} = \begin{bmatrix} 1.8 & 0.88 & 0.80 \\ 1 & 1.6 & 1 \\ 0.2 & 0.32 & 1.20 \end{bmatrix} \begin{bmatrix} 10 \\ 5 \\ 15 \end{bmatrix} = \begin{bmatrix} 34.4 \\ 33 \\ 21.6 \end{bmatrix}.$$

We conclude that $34.4 million worth of goods of one industry, $33 million worth of a second industry, and $21.6 million worth of a third industry should be produced.

USING TECHNOLOGY EXERCISES 2.7, page 173

1. The final outputs of the first, second, third, and fourth industries are 602.62, 502.30, 572.57, and 523.46 units, respectively.

3. The final outputs of the first, second, third, and fourth industries are 143.06, 132.98, 188.59, and 125.53 units, respectively.

CHAPTER 2, REVIEW EXERCISES, page 178

1. $\begin{bmatrix} 1 & 2 \\ -1 & 3 \\ 2 & 1 \end{bmatrix} + \begin{bmatrix} 1 & 0 \\ 0 & 1 \\ 1 & 2 \end{bmatrix} = \begin{bmatrix} 2 & 2 \\ -1 & 4 \\ 3 & 3 \end{bmatrix}.$ 3. $\begin{bmatrix} -3 & 2 & 1 \end{bmatrix} \begin{bmatrix} 2 & 1 \\ -1 & 0 \\ 2 & 1 \end{bmatrix} = \begin{bmatrix} -6 & -2 \end{bmatrix}.$

5. By the equality of matrices, $x = 2$, $z = 1$, $y = 3$ and $w = 3$.

7. By the equality of matrices,

$a + 3 = 6$, or $a = 3$.

$-1 = e + 2$, or $e = -3$; $b = 4$

$c + 1 = -1$, or $c = -2$; $d = 2$

$e + 2 = -1$, and $e = -3$.

9. $2A + 3B = 2\begin{bmatrix} 1 & 3 & 1 \\ -2 & 1 & 3 \\ 4 & 0 & 2 \end{bmatrix} + 3\begin{bmatrix} 2 & 1 & 3 \\ -2 & -1 & -1 \\ 1 & 4 & 2 \end{bmatrix} = \begin{bmatrix} 2 & 6 & 2 \\ -4 & 2 & 6 \\ 8 & 0 & 4 \end{bmatrix} + \begin{bmatrix} 6 & 3 & 9 \\ -6 & -3 & -3 \\ 3 & 12 & 6 \end{bmatrix}$

$= \begin{bmatrix} 8 & 9 & 11 \\ -10 & -1 & 3 \\ 11 & 12 & 10 \end{bmatrix}.$

11. $3A = \begin{bmatrix} 1 & 3 & 1 \\ -2 & 1 & 3 \\ 4 & 0 & 2 \end{bmatrix} = \begin{bmatrix} 3 & 9 & 3 \\ -6 & 3 & 9 \\ 12 & 0 & 6 \end{bmatrix}$

and $2(3A) = 2\begin{bmatrix} 3 & 9 & 3 \\ -6 & 3 & 9 \\ 12 & 0 & 6 \end{bmatrix} = \begin{bmatrix} 6 & 18 & 6 \\ -12 & 6 & 18 \\ 24 & 0 & 12 \end{bmatrix}.$

13. $B - C = \begin{bmatrix} 2 & 1 & 3 \\ -2 & -1 & -1 \\ 1 & 4 & 2 \end{bmatrix} - \begin{bmatrix} 3 & -1 & 2 \\ 1 & 6 & 4 \\ 2 & 1 & 3 \end{bmatrix} = \begin{bmatrix} -1 & 2 & 1 \\ -3 & -7 & -5 \\ -1 & 3 & -1 \end{bmatrix}$

and so $A(B - C) = \begin{bmatrix} 1 & 3 & 1 \\ -2 & 1 & 3 \\ 4 & 0 & 2 \end{bmatrix}\begin{bmatrix} -1 & 2 & 1 \\ -3 & -7 & -5 \\ -1 & 3 & -1 \end{bmatrix} = \begin{bmatrix} -11 & -16 & -15 \\ -4 & -2 & -10 \\ -6 & 14 & 2 \end{bmatrix}.$

15. $BC = \begin{bmatrix} 2 & 1 & 3 \\ -2 & -1 & -1 \\ 1 & 4 & 2 \end{bmatrix}\begin{bmatrix} 3 & -1 & 2 \\ 1 & 6 & 4 \\ 2 & 1 & 3 \end{bmatrix} = \begin{bmatrix} 13 & 7 & 17 \\ -9 & -5 & -11 \\ 11 & 25 & 24 \end{bmatrix}$

$ABC = \begin{bmatrix} 1 & 3 & 1 \\ -2 & 1 & 3 \\ 4 & 0 & 2 \end{bmatrix}\begin{bmatrix} 13 & 7 & 17 \\ -9 & -5 & -11 \\ 11 & 25 & 24 \end{bmatrix} = \begin{bmatrix} -3 & 17 & 8 \\ -2 & 56 & 27 \\ 74 & 78 & 116 \end{bmatrix}.$

17. Using the Gauss-Jordan elimination method, we find

$$\left[\begin{array}{cc|c} 2 & -3 & 5 \\ 3 & 4 & -1 \end{array}\right] \xrightarrow{\frac{1}{2}R_1} \left[\begin{array}{cc|c} 1 & -\frac{3}{2} & \frac{5}{2} \\ 3 & 4 & -1 \end{array}\right] \xrightarrow{R_2-3R_1} \left[\begin{array}{cc|c} 1 & -\frac{3}{2} & \frac{5}{2} \\ 0 & \frac{17}{2} & -\frac{17}{2} \end{array}\right]$$

$$\xrightarrow{\frac{2}{17}R_2} \left[\begin{array}{cc|c} 1 & -\frac{3}{2} & \frac{5}{2} \\ 0 & 1 & -1 \end{array}\right] \xrightarrow{R_1+\frac{3}{2}R_2} \left[\begin{array}{cc|c} 1 & 0 & 1 \\ 0 & 1 & -1 \end{array}\right]$$

We conclude that $x = 1$ and $y = -1$.

19.
$$\left[\begin{array}{ccc|c} 1 & -1 & 2 & 5 \\ 3 & 2 & 1 & 10 \\ 2 & -3 & -2 & -10 \end{array}\right] \xrightarrow[R_3-2R_1]{R_2-3R_1} \left[\begin{array}{ccc|c} 1 & -1 & 2 & 5 \\ 0 & 5 & -5 & -5 \\ 0 & -1 & -6 & -20 \end{array}\right] \xrightarrow{\frac{1}{5}R_2} \left[\begin{array}{ccc|c} 1 & -1 & 2 & 5 \\ 0 & 1 & -1 & -1 \\ 0 & -1 & -6 & -20 \end{array}\right]$$

$$\xrightarrow[R_3+R_2]{R_1+R_2} \left[\begin{array}{ccc|c} 1 & 0 & 1 & 4 \\ 0 & 1 & -1 & -1 \\ 0 & 0 & -7 & -21 \end{array}\right] \xrightarrow{-\frac{1}{7}R_3} \left[\begin{array}{ccc|c} 1 & 0 & 1 & 4 \\ 0 & 1 & -1 & -1 \\ 0 & 0 & 1 & 3 \end{array}\right] \xrightarrow[R_2+R_3]{R_1-R_3}$$

$$= \left[\begin{array}{ccc|c} 1 & 0 & 0 & 1 \\ 0 & 1 & 0 & 2 \\ 0 & 0 & 1 & 3 \end{array}\right]. \text{ Therefore, } x = 1, y = 2, \text{ and } z = 3.$$

21.
$$\left[\begin{array}{ccc|c} 3 & -2 & 4 & 11 \\ 2 & -4 & 5 & 4 \\ 1 & 2 & -1 & 10 \end{array}\right] \xrightarrow{R_1-R_2} \left[\begin{array}{ccc|c} 1 & 2 & -1 & 7 \\ 2 & -4 & 5 & 4 \\ 1 & 2 & -1 & 10 \end{array}\right] \xrightarrow[R_3-R_1]{R_2-2R_1} \left[\begin{array}{ccc|c} 1 & 2 & -1 & 7 \\ 0 & -8 & 7 & -10 \\ 0 & 0 & 0 & 3 \end{array}\right].$$

Since this last row implies that $0 = 3!$, we conclude that the system has no solution.

23.
$$\left[\begin{array}{ccc|c} 3 & -2 & 1 & 4 \\ 1 & 3 & -4 & -3 \\ 2 & -3 & 5 & 7 \\ 1 & -8 & 9 & 10 \end{array}\right] \xrightarrow{R_1-R_3} \left[\begin{array}{ccc|c} 1 & 1 & -4 & -3 \\ 1 & 3 & -4 & -3 \\ 2 & -3 & 5 & 7 \\ 1 & -8 & 9 & 10 \end{array}\right] \xrightarrow[R_4-R_1]{\substack{R_2-R_1 \\ R_3-2R_1}} \left[\begin{array}{ccc|c} 1 & 1 & -4 & -3 \\ 0 & 2 & 0 & 0 \\ 0 & -5 & 13 & 13 \\ 0 & -9 & 13 & 13 \end{array}\right]$$

$$\xrightarrow{\frac{1}{2}R_2} \begin{bmatrix} 1 & 1 & -4 & | & -3 \\ 0 & 1 & 0 & | & 0 \\ 0 & -5 & 13 & | & 13 \\ 0 & -9 & 13 & | & 13 \end{bmatrix} \xrightarrow[\substack{R_1-R_2 \\ R_3+5R_2 \\ R_4+9R_2}]{} \begin{bmatrix} 1 & 0 & -4 & | & -3 \\ 0 & 1 & 0 & | & 0 \\ 0 & 0 & 13 & | & 13 \\ 0 & 0 & 13 & | & 13 \end{bmatrix}$$

$$\xrightarrow{\frac{1}{13}R_3} \begin{bmatrix} 1 & 0 & -4 & | & -3 \\ 0 & 1 & 0 & | & 0 \\ 0 & 0 & 1 & | & 1 \\ 0 & 0 & 13 & | & 13 \end{bmatrix} \xrightarrow[\substack{R_1+4R_3 \\ R_4-13R_3}]{} \begin{bmatrix} 1 & 0 & 0 & | & 1 \\ 0 & 1 & 0 & | & 0 \\ 0 & 0 & 1 & | & 1 \\ 0 & 0 & 0 & | & 0 \end{bmatrix}.$$

Therefore, $x = 1$, $y = 0$, and $z = 1$.

25. $A^{-1} = \dfrac{1}{(3)(2)-(1)(1)} \begin{bmatrix} 2 & -1 \\ -1 & 3 \end{bmatrix} = \begin{bmatrix} \frac{2}{5} & -\frac{1}{5} \\ -\frac{1}{5} & \frac{3}{5} \end{bmatrix}.$

27. $A^{-1} = \dfrac{1}{(3)(2)-(2)(4)} \begin{bmatrix} 2 & -4 \\ -2 & 3 \end{bmatrix} = \begin{bmatrix} -1 & 2 \\ 1 & -\frac{3}{2} \end{bmatrix}.$

29. $\begin{bmatrix} 2 & 3 & 1 & | & 1 & 0 & 0 \\ 1 & -1 & 2 & | & 0 & 1 & 0 \\ 1 & 2 & 1 & | & 0 & 0 & 1 \end{bmatrix} \xrightarrow{R_1-R_2} \begin{bmatrix} 1 & 4 & -1 & | & 1 & -1 & 0 \\ 1 & -1 & 2 & | & 0 & 1 & 0 \\ 1 & 2 & 1 & | & 0 & 0 & 1 \end{bmatrix} \xrightarrow[\substack{R_2-R_1 \\ R_3-R_1}]{}$

$\begin{bmatrix} 1 & 4 & -1 & | & 1 & -1 & 0 \\ 0 & -5 & 3 & | & -1 & 2 & 0 \\ 0 & -2 & 2 & | & -1 & 1 & 1 \end{bmatrix} \xrightarrow{R_2-3R_3} \begin{bmatrix} 1 & 4 & -1 & | & 1 & -1 & 0 \\ 0 & 1 & -3 & | & 2 & -1 & -3 \\ 0 & -2 & 2 & | & -1 & 1 & 1 \end{bmatrix} \xrightarrow[\substack{R_1-4R_2 \\ R_3+2R_2}]{}$

$\begin{bmatrix} 1 & 0 & 11 & | & -7 & 3 & 12 \\ 0 & 1 & -3 & | & 2 & -1 & -3 \\ 0 & 0 & -4 & | & 3 & -1 & -5 \end{bmatrix} \xrightarrow{-\frac{1}{4}R_3} \begin{bmatrix} 1 & 0 & 11 & | & -7 & 3 & 12 \\ 0 & 1 & -3 & | & 2 & -1 & -3 \\ 0 & 0 & 1 & | & -\frac{3}{4} & \frac{1}{4} & \frac{5}{4} \end{bmatrix} \xrightarrow[\substack{R_1-11R_3 \\ R_2+3R_3}]{}$

$\begin{bmatrix} 1 & 0 & 0 & | & \frac{5}{4} & \frac{1}{4} & -\frac{7}{4} \\ 0 & 1 & 0 & | & -\frac{1}{4} & -\frac{1}{4} & \frac{3}{4} \\ 0 & 0 & 1 & | & -\frac{3}{4} & \frac{1}{4} & \frac{5}{4} \end{bmatrix}.$ So $A^{-1} = \begin{bmatrix} \frac{5}{4} & \frac{1}{4} & -\frac{7}{4} \\ -\frac{1}{4} & -\frac{1}{4} & \frac{3}{4} \\ -\frac{3}{4} & \frac{1}{4} & \frac{5}{4} \end{bmatrix}.$

31. $\begin{bmatrix} 1 & 2 & 4 & | & 1 & 0 & 0 \\ 3 & 1 & 2 & | & 0 & 1 & 0 \\ 1 & 0 & -6 & | & 0 & 0 & 1 \end{bmatrix} \xrightarrow[R_3-R_1]{R_2-3R_1} \begin{bmatrix} 1 & 2 & 4 & | & 1 & 0 & 0 \\ 0 & -5 & -10 & | & -3 & 1 & 0 \\ 0 & -2 & -10 & | & -1 & 0 & 1 \end{bmatrix} \xrightarrow{R_2-3R_3}$

$\begin{bmatrix} 1 & 2 & 4 & | & 1 & 0 & 0 \\ 0 & 1 & 20 & | & 0 & 1 & -3 \\ 0 & -2 & -10 & | & -1 & 0 & 1 \end{bmatrix} \xrightarrow[R_3+2R_2]{R_1-2R_2} \begin{bmatrix} 1 & 0 & -36 & | & 1 & -2 & 6 \\ 0 & 1 & 20 & | & 0 & 1 & -3 \\ 0 & 0 & 30 & | & -1 & 2 & -5 \end{bmatrix} \xrightarrow{\frac{1}{30}R_3}$

$\begin{bmatrix} 1 & 0 & -36 & | & 1 & -2 & 6 \\ 0 & 1 & 20 & | & 0 & 1 & -3 \\ 0 & 0 & 1 & | & -\frac{1}{30} & \frac{1}{15} & -\frac{1}{6} \end{bmatrix} \xrightarrow[R_2-20R_3]{R_1+36R_3} \begin{bmatrix} 1 & 0 & 0 & | & -\frac{1}{5} & \frac{2}{5} & 0 \\ 0 & 1 & 0 & | & \frac{2}{3} & -\frac{1}{3} & \frac{1}{3} \\ 0 & 0 & 1 & | & -\frac{1}{30} & \frac{1}{15} & -\frac{1}{6} \end{bmatrix}$

So $\qquad A^{-1} = \begin{bmatrix} -\frac{1}{5} & \frac{2}{5} & 0 \\ \frac{2}{3} & -\frac{1}{3} & \frac{1}{3} \\ -\frac{1}{30} & \frac{1}{15} & -\frac{1}{6} \end{bmatrix}.$

33. $(A^{-1}B)^{-1} = B^{-1}(A^{-1})^{-1} = B^{-1}A$. Now

$$B^{-1} = \frac{1}{(3)(2)-4(1)}\begin{bmatrix} 2 & -1 \\ -4 & 3 \end{bmatrix} = \begin{bmatrix} 1 & -\frac{1}{2} \\ -2 & \frac{3}{2} \end{bmatrix}.$$

$$B^{-1}A = \begin{bmatrix} 1 & -\frac{1}{2} \\ -2 & \frac{3}{2} \end{bmatrix}\begin{bmatrix} 1 & 2 \\ -1 & 2 \end{bmatrix} = \begin{bmatrix} \frac{3}{2} & 1 \\ -\frac{7}{2} & -1 \end{bmatrix}.$$

35. $2A - C = \begin{bmatrix} 2 & 4 \\ -2 & 4 \end{bmatrix} - \begin{bmatrix} 1 & 1 \\ -1 & 2 \end{bmatrix} = \begin{bmatrix} 1 & 3 \\ -1 & 2 \end{bmatrix}.$

$$(2A-C)^{-1} = \frac{1}{(1)(2)-(-1)(3)}\begin{bmatrix} 2 & -3 \\ 1 & 1 \end{bmatrix} = \begin{bmatrix} \frac{2}{5} & -\frac{3}{5} \\ \frac{1}{5} & \frac{1}{5} \end{bmatrix}.$$

37. $\qquad A = \begin{bmatrix} 2 & 3 \\ 1 & -2 \end{bmatrix}, \quad X = \begin{bmatrix} x \\ y \end{bmatrix}, \quad C = \begin{bmatrix} -8 \\ 3 \end{bmatrix}$

$$A^{-1} = \frac{1}{(-2)(2)-(1)(3)}\begin{bmatrix} -2 & -3 \\ -1 & 2 \end{bmatrix} = \begin{bmatrix} \frac{2}{7} & \frac{3}{7} \\ \frac{1}{7} & -\frac{2}{7} \end{bmatrix}$$

$$\begin{bmatrix} x \\ y \end{bmatrix} = A^{-1}B = \begin{bmatrix} \frac{2}{7} & \frac{3}{7} \\ \frac{1}{7} & -\frac{2}{7} \end{bmatrix}\begin{bmatrix} -8 \\ 3 \end{bmatrix} = \begin{bmatrix} -1 \\ -2 \end{bmatrix}.$$

39. Put

$$X = \begin{bmatrix} x \\ y \\ z \end{bmatrix}, \quad A = \begin{bmatrix} 1 & -2 & 4 \\ 2 & 3 & -2 \\ 1 & 4 & -6 \end{bmatrix}, \quad C = \begin{bmatrix} 13 \\ 0 \\ -15 \end{bmatrix}.$$

Then $AX = C$ and $X = A^{-1}C$. To find A^{-1},

$$\left[\begin{array}{ccc|ccc} 1 & -2 & 4 & 1 & 0 & 0 \\ 2 & 3 & -2 & 0 & 1 & 0 \\ 1 & 4 & -6 & 0 & 0 & 1 \end{array}\right] \xrightarrow[R_3-R_1]{R_2-2R_1} \left[\begin{array}{ccc|ccc} 1 & -2 & 4 & 1 & 0 & 0 \\ 0 & 7 & -10 & -2 & 1 & 0 \\ 0 & 6 & -10 & -1 & 0 & 1 \end{array}\right] \xrightarrow{R_2-R_3}$$

$$\left[\begin{array}{ccc|ccc} 1 & -2 & 4 & 1 & 0 & 0 \\ 0 & 1 & 0 & -1 & 1 & -1 \\ 0 & 6 & -10 & -1 & 0 & 1 \end{array}\right] \xrightarrow[R_3-6R_2]{R_1+2R_2} \left[\begin{array}{ccc|ccc} 1 & 0 & 4 & -1 & 2 & -2 \\ 0 & 1 & 0 & -1 & 1 & -1 \\ 0 & 0 & -10 & 5 & -6 & 7 \end{array}\right] \xrightarrow{-\frac{1}{10}R_3}$$

$$\left[\begin{array}{ccc|ccc} 1 & 0 & 4 & -1 & 2 & -2 \\ 0 & 1 & 0 & -1 & 1 & -1 \\ 0 & 0 & 1 & -\frac{1}{2} & \frac{3}{5} & -\frac{7}{10} \end{array}\right] \xrightarrow{R_1-4R_3} \left[\begin{array}{ccc|ccc} 1 & 0 & 0 & 1 & -\frac{2}{5} & \frac{4}{5} \\ 0 & 1 & 0 & -1 & 1 & -1 \\ 0 & 0 & 1 & -\frac{1}{2} & \frac{3}{5} & -\frac{7}{10} \end{array}\right].$$

So $\quad A^{-1} = \begin{bmatrix} 1 & -\frac{2}{5} & \frac{4}{5} \\ -1 & 1 & -1 \\ -\frac{1}{2} & \frac{3}{5} & -\frac{7}{10} \end{bmatrix}.$

Therefore, $\quad X = A^{-1}C = \begin{bmatrix} 1 & -\frac{2}{5} & \frac{4}{5} \\ -1 & 1 & -1 \\ -\frac{1}{2} & \frac{3}{5} & -\frac{7}{10} \end{bmatrix}\begin{bmatrix} 13 \\ 0 \\ -15 \end{bmatrix} = \begin{bmatrix} 1 \\ 2 \\ 4 \end{bmatrix}$

that is, $x = 1$, $y = 2$, and $z = 4$.

2 Systems of Linear Equations and Matrices

$$41. \quad \begin{bmatrix} x \\ y \\ z \end{bmatrix} = \begin{bmatrix} 600 & 1000 & 800 & 1400 \\ 700 & 800 & 600 & 1200 \\ 1200 & 800 & 1000 & 900 \end{bmatrix} \begin{bmatrix} 1.60 \\ 1.20 \\ 1.50 \\ 1.30 \end{bmatrix} = \begin{bmatrix} 5180 \\ 4540 \\ 5550 \end{bmatrix}.$$

The total revenue is \$5180 at station A, \$4540 at station B, and \$5550 at station C.

43. We wish to solve the system of equations

$$2x + 2y + 3z = 210$$
$$2x + 3y + 4z = 270$$
$$3x + 4y + 3z = 300.$$

Using the Gauss–Jordan method of elimination, we find

$$\begin{bmatrix} 2 & 2 & 3 & | & 210 \\ 2 & 3 & 4 & | & 270 \\ 3 & 4 & 3 & | & 300 \end{bmatrix} \xrightarrow{\frac{1}{2}R_1} \begin{bmatrix} 1 & 1 & \frac{3}{2} & | & 105 \\ 2 & 3 & 4 & | & 270 \\ 3 & 4 & 3 & | & 300 \end{bmatrix} \xrightarrow[R_3 - 3R_1]{R_2 - 2R_1} \begin{bmatrix} 1 & 1 & \frac{3}{2} & | & 105 \\ 0 & 1 & 1 & | & 60 \\ 0 & 1 & -\frac{3}{2} & | & -15 \end{bmatrix}$$

$$\xrightarrow[R_3 - R_2]{R_1 - R_2} \begin{bmatrix} 1 & 0 & \frac{1}{2} & | & 45 \\ 0 & 1 & 1 & | & 60 \\ 0 & 0 & -\frac{5}{2} & | & -75 \end{bmatrix} \xrightarrow{-\frac{2}{5}R_3} \begin{bmatrix} 1 & 0 & \frac{1}{2} & | & 45 \\ 0 & 1 & 1 & | & 60 \\ 0 & 0 & 1 & | & 30 \end{bmatrix} \xrightarrow[R_2 - R_3]{R_1 - \frac{1}{2}R_3} \begin{bmatrix} 1 & 0 & 0 & | & 30 \\ 0 & 1 & 0 & | & 30 \\ 0 & 0 & 1 & | & 30 \end{bmatrix}$$

So $x = y = z = 30$. Therefore, Desmond should produce 30 of each type of pendant.

CHAPTER 3

EXERCISES 3.1, page 188

1. $4x - 8 < 0$ implies $x < 2$. The graph of the inequality is at the right.

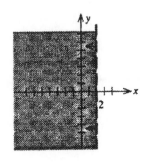

3. $x - y \leq 0$ implies $x \leq y$. The graph of the inequality is at the right.

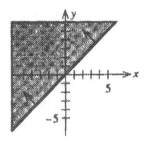

5. The graph of the inequality $x \leq -3$ is at the right.

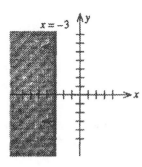

7. We first sketch the straight line with
equation $2x + y = 4$. Next, picking
the test point $(0,0)$, we have
$$2(0) + (0) = 0 \leq 4.$$
We conclude that the half-plane
containing the origin is the required half-plane.

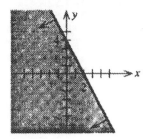

9. We first sketch the graph of the straight
line
$$4x - 3y = -24.$$

Next, picking the test point $(0,0)$, we see that
$$4(0) \ - 3(0) = 0 \not< -24.$$

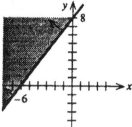

We conclude that the half-plane not
containing the origin is the required half-plane. The graph of this inequality is at the
right.

11. The system of linear inequalities that describes the shaded region is
$$x \geq 1, x \leq 5, y \geq 2, \text{ and } y \leq 4.$$

We may also combine the first and second inequalities and the third and fourth
inequalities and write
$$1 \leq x \leq 5 \quad \text{and} \quad 2 \leq y \leq 4.$$

13. The system of linear inequalities that describes the shaded region is
$$2x - y \geq 2, 5x + 7y \geq 35, \text{ and } x \leq 4.$$

15. The system of linear inequalities that describes the shaded region is
$$7x + 4y \leq 140, x + 3y \geq 30, \text{ and } x \ - y \geq \ -10.$$

17. The system of linear inequalities that describes the shaded region is
$$x + y \geq 7, x \geq 2, y \geq 3, \text{ and } y \leq 7.$$

19. The required solution set is shown at the
    right. To find the coordinates of A, we
    solve the system

    $$2x + 4y = 16$$
    $$-x + 3y = 7,$$

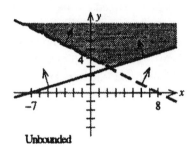

Unbounded

    giving $A = (2,3)$. Observe that a dotted line
    is used to show that no point on the line
    constitutes a solution to the given problem.
    Observe that this is an unbounded solution
    set.

21. The solution set is shown in the figure at
    the right. Observe that the set is unbounded.
    To find the coordinates of A, we solve the
    system

    $$x - y = 0$$
    $$2x + 3y = 10$$

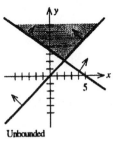

Unbounded

    giving $A = (2,2)$. Observe that this is an unbounded
    solution set.

23. The half-planes defined by the two inequalities are
    shown in the figure at the right. Since the
    two half-planes have no points in common, we
    conclude that the given system of inequalities has
    no solution. (The empty set is a bounded set.)

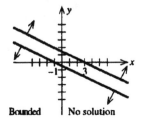

Bounded   No solution

25. The half-planes defined by the three inequalities
    are shown in the figure at the right. The point A is
    found by solving the system

    $$x + y = 6$$
    $$x = 3$$

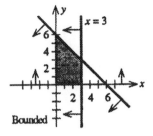

Bounded

    giving $A = (3,3)$. Observe that this is a bounded
    solution set.

27. The half-planes defined by the given inequalities are shown in the figure at the right. Observe that the two lines described by the equations
$$3x - 6y = 12 \quad \text{and} \quad -x + 2y = 4$$
do not intersect because they are parallel. The solution set is unbounded.

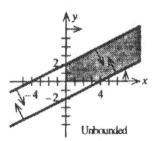

29. The required solution set is shown in the figure at the right. The coordinates of A are found by solving the system
$$3x - 7y = -24$$
$$x + 3y = 8$$
giving $(-1,3)$.
The solution set is unbounded.

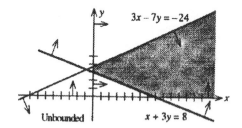

31. The required solution set is shown in the figure at the right. The solution set is bounded.

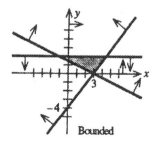

33. The required solution set is shown in the figure at the right. The solution set has vertices at $(0,6)$, $(5,0)$, $(4,0)$, and $(1,3)$. The solution set is bounded.

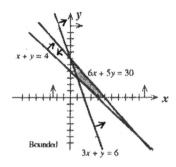

35. The required solution set is shown in the figure at the right. The unbounded solution set has vertices at (2,8), (0,6), (0,3),and (2,2).

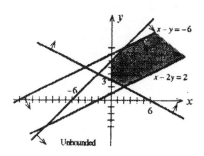

37. False. It is always a half-plane. A straight line is the graph of a linear equation and vice-versa.

39. True. Since a circle can always be enclosed by a rectangle, the solution set of such a system is bounded if it can be enclosed by a rectangle.

EXERCISES 3.2, page 188

1. We tabulate the given information:

	Product A	Product B	Time available
Machine I	6	9	300
Machine II	5	4	180
Profit per unit ($)	3	4	

Let x and y denote the number of units of Product A and Product B to be produced. Then the required linear programming problem is:
Maximize $P = 3x + 4y$ subject to the constraints
$$6x + 9y \le 300$$
$$5x + 4y \le 180$$
$$x \ge 0, y \ge 0$$

3. Let x denote the number of model A grates to be produced and y denote the number of model B grates to be produced. Since only 1000 pounds of cast iron are available, we must have
$$3x + 4y \le 1000.$$

The restriction that only 20 hours of labor are available per day implies that
$$6x + 3y \leq 1200. \quad \text{(time in minutes)}$$
Then the profit on the production of these grates is given by
$$P = 2x + 1.5y.$$
The additional restriction that at least 150 model A grates be produced each day implies that
$$x \geq 150.$$
Summarizing, we have the following linear programming problem:

Maximize $P = 2x + 1.5y$ subject to
$$3x + 4y \leq 1000$$
$$6x + 3y \leq 1200$$
$$x \geq 150, y \geq 0$$

5.   Let x denote the number of fully assembled units to be produced daily and let y denote the number of kits to be produced. Then the fraction of the day the fabrication department works on the fully assembled cabinets is $\frac{1}{200}x$. Similarly the fraction of the day the fabrication department works on kits is $\frac{1}{200}y$. Since the fraction of the day during which the fabrication department is busy cannot exceed one, we must have
$$\frac{1}{200}x + \frac{1}{200}y \leq 1.$$
Similarly, the restrictions place on the assembly department leads to the inequality
$$\frac{1}{100}x + \frac{1}{300}y \leq 1.$$
The profit (objective) function is $P = 50x + 40y$. Summarizing, the required linear programming problem is

Maximize $P = 50x + 40y$ subject to
$$\frac{1}{200}x + \frac{1}{200}y \leq 1$$
$$\frac{1}{100}x + \frac{1}{300}y \leq 1$$
$$x \geq 0, \ y \geq 0.$$

7.   Let x and y denote the amount of food A and food B, respectively, used to prepare a meal. Then the requirement that the meal contain a minimum of 400 mg of calcium implies
$$30x + 25y \geq 400.$$

Similarly, the requirements that the meal contain at least 10 mg of iron and 40 mg of vitamin C imply that

$$x + 0.5y \geq 10$$

$$2x + 5y \geq 40.$$

The cholesterol content is given by

$$C = 2x + 5y.$$

Therefore, the linear programming problem is

Minimize $C = 2x + 5y$ subject to

$$30x + 25y \geq 400$$

$$x + 0.5y \geq 10$$

$$2x + 5y \geq 40$$

$$x \geq 0, y \geq 0$$

9. Let x denote the number of picture tubes shipped from location I to city A and let y denote the number of picture tubes shipped from location I to city B. Since the number of picture tubes required by the two factories in city A and city B are 3000 and 4000, respectively, the number of picture tubes shipped from location II to city A and city B, are (3000 - x) and (4000 - y), respectively. These numbers are shown in the following schematic,

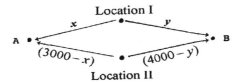

Referring to the schematic and the shipping schedule, we find that the total shipping costs incurred by the company are given by

$$C = 3x + 2y + 4(3000 - x) + 5(4000 - y)$$

$$= 32000 - x - 3y$$

The production constraints on Location I and II lead to the inequalities

$$x + \quad\quad y \leq 6000$$

$$(3000 - x) + (4000 - y) \leq 5000$$

This last inequality simplifies to

$$x + y \geq 2000.$$

The requirements of the two factories lead to the inequalities

$$x \geq 0, y \geq 0, 3000 - x \geq 0, \text{ and } 4000 - y \geq 0.$$

These last two inequalities may be written as $x \le 3000$ and $y \le 4000$.
Summarizing, we have the following linear programming problem:

Minimize $C = 32{,}000 - x - 3y$ subject to

$$x + y \le 6000$$
$$x + y \ge 2000$$
$$x \le 3000$$
$$y \le 4000$$
$$x \ge 0, \ y \ge 0$$

11. Let x, y, and z denote the number of units produced of products A, B, and C, respectively. From the given information, we formulate the following linear programming problem:

Maximize $P = 18x + 12y + 15z$ subject to

$$2x + y + 2z \le 900$$
$$3x + y + 2z \le 1080$$
$$2x + 2y + z \le 840$$
$$x \ge 0, y \ge 0, z \ge 0$$

13. We first tabulate the given information:

Dept.	MODELS			
	A	B	C	Time available
Fabrication	$\frac{5}{4}$	$\frac{3}{2}$	$\frac{3}{2}$	310
Assembly	1	1	$\frac{3}{4}$	205
Finishing	1	1	$\frac{1}{2}$	190

Let x, y, and z denote the number of units of model A, model B, and model C to be produced, respectively. Then the required linear programming problem is

Maximize $P = 26x + 28y + 24z$ subject to
$$\frac{5}{4}x + \frac{3}{2}y + \frac{3}{2}z \leq 310$$
$$x + y + \frac{3}{4}z \leq 205$$
$$x + y + \frac{1}{2}z \leq 190$$
$$x \geq 0, y \geq 0, z \geq 0$$

15. The shipping costs are tabulated in the following table.

	Warehouse A	Warehouse B	Warehouse C
Plant I	60	60	80
Plant II	80	70	50

Letting x_1 denote the number of pianos shipped from plant I to warehouse A, x_2 the number of pianos shipped from plant I to warehouse B, and so we have

	Warehouse			Max. Production
	A	B	C	
Plant I	x_1	x_2	x_3	300
Plant II	x_4	x_5	x_6	250
Min. Req.	200	150	200	

From the two tables we see that the total monthly shipping cost is given by

$$C = 60x_1 + 60x_2 + 80x_3 + 80x_4 + 70x_5 + 50x_6.$$

Next, the production constraints on plants I and II lead to the inequalities
$$x_1 + x_4 \geq 200$$
$$x_2 + x_5 \geq 150$$
$$x_3 + x_6 \geq 200$$

Summarizing we have the following linear programming problem:

Minimize $C = 60x_1 + 60x_2 + 80x_3 + 80x_4 + 70x_5 + 50x_6$ subject to

$$x_1 + x_2 + x_3 \le 300$$
$$x_4 + x_5 + x_6 \le 250$$
$$x_1 + x_4 \ge 200$$
$$x_2 + x_5 \ge 150$$
$$x_3 + x_6 \ge 200$$
$$x_1 \ge 0, \ x_2 \ge 0, \ ..., \ x_6 \ge 0$$

17. False. The objective function $P = xy$ is not a linear function in x and y.

EXERCISES 3.3, page 210

1. Evaluating the objective function at each of the corner points we obtain the following table.

Vertex	$Z = 2x + 3y$
(1,1)	5
(8,5)	31
(4,9)	35
(2,8)	28

From the table, we conclude that the maximum value of Z is 35 and it occurs at the vertex (4,9). The minimum value of Z is 5 and it occurs at the vertex (1,1).

3. Evaluating the objective function at each of the corner points we obtain the following table.

Vertex	$Z = 3x + 4y$
(0,20)	80
(3,10)	49
(4,6)	36
(9,0)	27

From the graph, we conclude that there is no maximum value since Z is unbounded. The minimum value of Z is 27 and it occurs at the vertex (9,0).

5. Evaluating the objective function at each of the corner points we obtain the

following table.

Vertex	$Z = x + 4y$
(0,6)	24
(4,10)	44
(12,8)	44
(15,0)	15

From the table, we conclude that the maximum value of Z is 44 and it occurs at every point on the line segment joining the points (4,10) and (12,8). The minimum value of Z is 15 and it occurs at the vertex (15,0).

7. The problem is to maximize $P = 2x + 3y$ subject to
$$x + y \leq 6$$
$$x \leq 3$$
$$x \geq 0, y \geq 0$$

The feasible set S for the problem is shown in the following figure, and the values of the function P at the vertices of S are summarized in the accompanying table.

Vertex	$P = 2x + 3y$
$A(0,0)$	0
$B(3,0)$	6
$C(3,3)$	15
$D(0,6)$	18

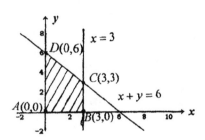

We conclude that P attains a maximum value of 18 when $x = 0$ and $y = 6$.

9. The problem is

Minimize $C = 2x + 10y$ subject to
$$5x + 2y \geq 40$$
$$x + 2y \geq 20$$
$$y \geq 3, \ x \geq 0$$

The feasible set S for the problem is shown in the figure that follows and the values of the function C at the vertices of S are summarized in the accompanying table.

Vertex	$C = 2x + 10y$
$A(0,20)$	200
$B(5,\frac{15}{2})$	85
$C(14,3)$	58

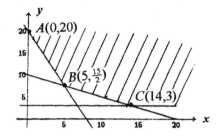

We conclude that C attains a minimum value of 58 when $x = 14$ and $y = 3$.

11. The problem is to minimize $C = 6x + 3y$ subject to

$$4x + y \geq 40$$
$$2x + y \geq 30$$
$$x + 3y \geq 30$$
$$x \geq 0, y \geq 0$$

The feasible set S is shown in the following figure, and the values of C at each of the vertices of S are shown in the accompanying table.

Vertex	$C = 6x + 3y$
$A(0,40)$	120
$B(5,20)$	90
$C(12,6)$	90
$D(30,0)$	180

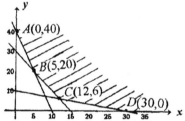

We conclude that C attains a minimum value of 90 at any point (x, y) lying on the line segment joining $(5,20)$ to $(12,6)$.

13. The problem is to minimize $C = 10x + 15y$ subject to

$$x + y \leq 10$$
$$3x + y \geq 12$$
$$-2x + 3y \geq 3$$
$$x \geq 0, y \geq 0$$

The feasible set is shown in the following figure, and the values of C at each of the vertices of S are shown in the accompanying table.

Vertex	$C = 10x + 15y$
$A(3,3)$	75
$B(\frac{27}{5}, \frac{23}{5})$	123
$C(1,9)$	145

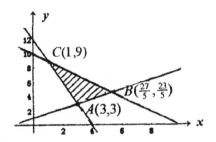

We conclude that C attains a minimum value of 75 when $x = 3$ and $y = 3$.

15. The problem is to maximize $P = 3x + 4y$ subject to

$$x + 2y \le 50$$
$$5x + 4y \le 145$$
$$2x + y \ge 25$$
$$y \ge 5, x \ge 0$$

The feasible set S is shown in the figure that follows, and the values of P at each of the vertices of S are shown in the accompanying table.

Vertex	$P = 3x + 4y$
$A(10,5)$	50
$B(25,5)$	95
$C(15, \frac{35}{2})$	115
$D(0,25)$	100

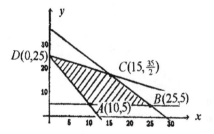

We conclude that P attains a maximum value of 115 when $x = 15$ and $y = 35/2$.

17. The problem is to maximize $P = 2x + 3y$ subject to

$$x + y \le 48$$
$$x + 3y \ge 60$$
$$9x + 5y \le 320$$
$$x \ge 10, y \ge 0$$

The feasible set S is shown in the figure that follows, and the values of P at each of the vertices of S are shown in the accompanying table.

Vertex	$P = 2x + 3y$
$A(10, \frac{50}{3})$	70
$B(30,10)$	90
$C(20,28)$	124
$D(10,38)$	134

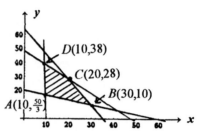

We conclude that P attains a maximum value of 134 when $x = 10$ and $y = 38$.

19. The problem is to find the maximum and minimum value of $P = 10x + 12y$ subject to

$$5x + 2y \geq 63$$
$$x + y \geq 18$$
$$3x + 2y \leq 51$$
$$x \geq 0, y \geq 0$$

The feasible set is shown at the right and the value of P at each of the vertices of S are shown in the accompanying table.

Vertex	$P = 10x + 12y$
$A(9,9)$	198
$B(15,3)$	186
$C(6, \frac{33}{2})$	258

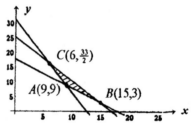

P attains a maximum value of 258 when $x = 6$ and $y = 33/2$. The minimum value of P is 186. It is attained when $x = 15$ and $y = 3$.

21. The problem is to find the maximum and minimum value of $P = 2x + 4y$ subject to

$$x + y \leq 20$$
$$-x + y \leq 10$$
$$x \leq 10$$
$$x + y \geq 5$$
$$y \geq 5, \quad x \geq 0$$

The feasible set is shown in the figure that follows, and the value of P at each of the vertices of S are shown in the accompanying table.

Vertex	$P = 2x + 4y$
$A(0,5)$	20
$B(10,5)$	40
$C(10,10)$	60
$D(5,15)$	70
$E(0,10)$	40

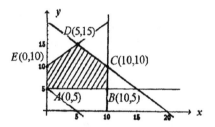

P attains a maximum value of 70 when $x = 5$ and $y = 15$. The minimum value of P is 20. It is attained when $x = 0$ and $y = 5$.

23. Let x and y denote the number of model A and model B fax machines produced in each shift. Then the restriction on manufacturing costs implies
$$200x + 300y \leq 600,000,$$
and the limitation on the number produced implies
$$x + y \leq 2,500.$$
The total profit is $P = 25x + 40y$. Summarizing, we have the following linear programming problem.
    Maximize $P = 25x + 40y$ subject to

$$200x + 300y \leq 600,000$$
$$x + \quad y \leq \quad 2,500$$
$$x \geq 0, \ y \geq 0$$

The graph of the feasible set S and the associated table of values of P follow.

Vertex	$C = 25x + 40y$
$A(0,0)$	0
$B(2500,0)$	62,500
$C(1500,1000)$	77,500
$D(0,2000)$	80,000

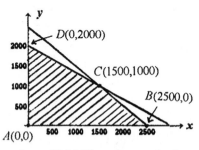

P attains a maximum value of 80,000 when $x = 0$ and $y = 2000$. Thus, by producing 2000 model B fax machines in each shift, the company will realize an optimal profit of $80,000.

25. Refer to the solution of Exercise 2, Section 3.2, The problem is
Maximize $P = 2x + 1.5y$ subject to
$$3x + 4y \le 1000$$
$$6x + 3y \le 1200$$
$$x \ge 0, \ y \ge 0$$

The graph of the feasible set S and the associated table of values of P follow.

Vertex	$P = 2x + 1.5y$
$A(0,0)$	0
$B(200,0)$	400
$C(120,160)$	480
$D(0,250)$	375

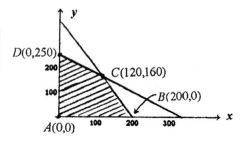

P attains a maximum value of 480 when $x = 120$ and $y = 160$. Thus, by producing 120 model A grates and 160 model B grates in each shift, the company will realize an optimal profit of $480.

27. Refer to the solution of Exercise 4, Section 3.2. The linear programming problem is
Maximize $P = 0.1x + 0.12y$ subject to
$$x + y \le 20$$
$$x - 4y \ge 0$$
$$x \ge 0, \ y \ge 0$$

The feasible set S for the problem is shown in the figure at the right, and the value of P at each of the vertices of S is shown in the accompanying table.

Vertex	$C = 0.1x + 0.12y$
$A(0,0)$	0
$B(16,4)$	2.08
$C(20,0)$	2.00

The maximum value of P is attained when $x = 16$ and $y = 4$. Thus, by extending $16 million in housing loans and $4 million in automobile loans, the company will realize a return of $2.08 million on its loans.

29. Refer to Exercise 6, Section 3.2. The problem is

Maximize $P = 150x + 200y$ subject to

$$40x + 60y \leq 7400$$

$$20x + 25y \leq 3300$$

$$x \geq 0, \; y \geq 0$$

The graph of the feasible set S and the associated table of values of P follow.

Vertex	$P = 150x + 200y$
$A(0,0)$	0
$B(165,0)$	24,750
$C(65,80)$	25,750
$D(0,123)$	24,600

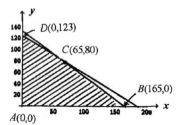

P attains a maximum value of 25,750 when $x = 65$ and $y = 80$. Thus, by producing 65 acres of crop A and 80 acres of crop B, the farmer will realize a maximum profit of $25,750.

31. The problem is

Minimize $C = 14,500 - 20x - 10y$ subject to

$$x + y \geq 40$$

$$x + y \leq 100$$

$$0 \leq x \leq 80$$

$$0 \leq y \leq 70$$

The feasible set S for the problem is shown in the figure at the right, and the value of C at each of the vertices of S is given in the accompanying table.

Vertex	$C = 14,500 - 20x - 10y$
$A(40,0)$	13,700
$B(80,0)$	12,900
$C(80,20)$	12,700
$D(30,70)$	13,200
$E(0,70)$	13,800
$F(0,40)$	14,100

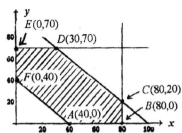

We conclude that the minimum value of C occurs when $x = 80$ and $y = 20$. Thus, 80 engines should be shipped from plant I to assembly plant A, and 20 engines should be shipped from plant I to assembly plant B; whereas

$(80 - x) = 80 - 80 = 0,$ and $(70 - y) = 70 - 20 = 50$
engines should be shipped from plant II to assembly plants A and B, respectively, at a total cost of $12,700.

33. Refer to the solution of Exercise 9, Section 3.2.

Minimize $C = 32,000 - x - 3y$ subject to
$$x + y \leq 6000$$
$$x + y \geq 2000$$
$$x \leq 3000$$
$$y \leq 4000$$
$$x \geq 0, \ y \geq 0$$

Vertex	$C = 32,000 - x - 3y$
$A(2000,0)$	30,000
$B(3000,0)$	29,000
$C(3000,3000)$	20,000
$D(2000,4000)$	18,000
$E(0,4000)$	20,000
$F(0,2000)$	26,000

Since x denotes the number of picture tubes shipped from location I to city A and y denotes the number of picture tubes shipped from location I to city B, we see that the company should ship 2000 tubes from location I to city A and 4000 tubes from location I to city B. Since the number of picture tubes required by the two factories in city A and city B are 3000 and 4000, respectively, the number of picture tubes shipped from location II to city A and city B, are

$(3000 - x) = 3000 - 2000 = 1000$ and $(4000 - y) = 4000 - 4000 = 0$

respectively. The minimum shipping cost will then be $18,000.

35. Let x denote Patricia's investment in growth stocks and y denote the value of her investment in speculative stocks, where both x and y are measured in thousands of dollars. Then the return on her investments is given by $P = 0.15x + 0.25y$. Since her

investment may not exceed \$30,000, we have the constraint $x + y \leq 30$. The condition that her investment in growth stocks be at least 3 times as much as her investment in speculative stocks translates into the inequality $x \geq 3y$. Thus, we have the following linear programming problem:

Maximize $P = 0.15x + 0.25y$ subject to

$$x + y \leq 30$$

$$x - 3y \geq 0$$

$$x \geq 0, \ y \geq 0$$

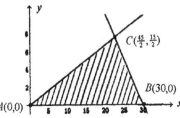

Vertex	$C = 0.15x + 0.25y$
$A(0,0)$	0
$B(30,0)$	4.5
$C(\frac{45}{2}, \frac{15}{2})$	5.25

The graph of the feasible set S is shown in the figure and the value of P at each of the vertices of S is shown in the accompanying table. The maximum value of P occurs when $x = 22.5$ and $y = 7.5$. Thus, by investing \$22,500 in growth stocks and \$7,500 in speculative stocks, Patricia will realize a return of \$5250 on her investments.

37. True. The optimal solution of a linear programming problem is the largest of all the feasible solutions.

39. False. The feasible set could be empty, and therefore, bounded.

41. Since the point $Q(x_1, y_1)$ lies in the interior of the feasible set S, it is possible to find another point $P(x_1, y_2)$ lying to the right and above the point Q and contained in S. (See figure.) Clearly, $x_2 > x_1$ and $y_2 > y_1$. Therefore, $ax_2 + by_2 > ax_1 + by_1$, since $a > 0$ and $b > 0$ and this shows that the objective function $P = ax + by$ takes on a larger value at P than it does at Q. Therefore, the optimal solution cannot occur at Q.

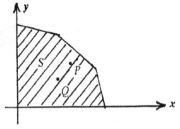

43. a.

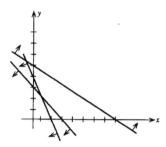

b. There is no point that satisfies all the given inequalities. Therefore, there is no solution.

CHAPTER 3 REVIEW EXERCISES, page 216

1. Evaluating Z at each of the corners of the feasible set S, we obtain the following table.

Vertex	$Z = 2x + 3y$
(0,0)	0
(5,0)	10
(3,4)	18
(0,6)	18

We conclude that Z attains a minimum value of 0 when $x = 0$ and $y = 0$, and a maximum value of 18 when x and y lie on the line segment joining (3,4) and (0,6).

3. The graph of the feasible set S is shown at the right.

Vertex	$Z = 3x + 5y$
$A(0,0)$	0
$B(5,0)$	15
$C(3,2)$	19
$D(0,4)$	20

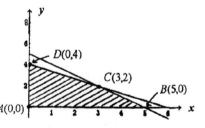

We conclude that the maximum value of P is 20 when $x = 0$ and $y = 4$.

5. The values of the objective function $C = 2x + 5y$ at the corners of the feasible set are given in the following table. The graph of the feasible set S follows.

Vertex	$C = 2x + 5y$
$A(0,0)$	0
$B(4,0)$	8
$C(3,4)$	26
$D(0,5)$	25

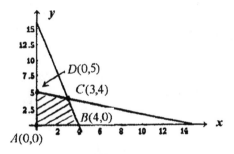

We conclude that the minimum value of C is 0 when $x = 0$ and $y = 0$.

7. The values of the objective function $P = 3x + 2y$ at the vertices of the feasible set are given in the following table. The graph of the feasible set is shown at the right.

Vertex	$P = 3x + 2y$
$A(0, \frac{28}{5})$	$\frac{56}{5}$
$B(7,0)$	21
$C(8,0)$	24
$D(3,10)$	29
$E(0,12)$	24

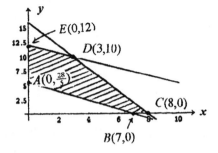

We conclude that P attains a maximum value of 29 when $x = 3$ and $y = 10$.

9. The graph of the feasible set S is shown at the right. The values of the objective function $C = 2x + 7y$ at each of the corner points of the feasible set S are shown in the table that follows.

Vertex	$C = 2x + 7y$
$A(20,0)$	40
$B(10,3)$	41
$C(0,9)$	63

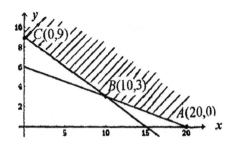

We conclude that C attains a minimum value of 40 when $x = 20$ and $y = 0$.

11. The graph of the feasible set S is shown in the following figure. We conclude that Q

attains a maximum value of 22 when $x = 22$ and $y = 0$, and a minimum value of $5\frac{1}{2}$ when $x = 3$ and $y = \frac{5}{2}$.

Vertex	$Q = x + y$
$A(2,5)$	7
$B(3,\frac{5}{2})$	$5\frac{1}{2}$
$C(8,0)$	8
$D(22,0)$	22

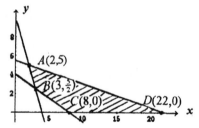

13. Suppose the investor puts x and y thousand dollars into the stocks of company A and company B, respectively. Then the mathematical formulation leads to the linear programming problem:

Maximize $P = 0.14x + 0.20y$ subject to
$$x + \quad y \le 80$$
$$0.01x + 0.04y \le 2$$
$$x \ge 0, \; y \ge 0$$

The feasible set S for this problem is shown in figure at the right and the values at each corner point are given in the accompanying table.

Vertex	$P = 0.14x + 0.20y$
$A(0,0)$	0
$B(80,0)$	11.2
$C(40,40)$	13.6
$D(0,50)$	10

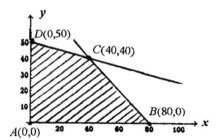

P attains a maximum value of 13.6 when $x = 40$ and $y = 40$. Thus, by investing $40,000 in the stocks of each company, the investor will achieve a maximum return of $13,600.

15. Let x denote the number of model A hibachis and y the number of required is
$$3x + 4y \le 1000$$
and the number of minutes of labor used each day is
$$6x + 3y \le 1200.$$
One additional constraint specifies that $y \ge 180$. The daily profit is $P = 2x + 1.5y$.

Therefore, we have the following linear programming problem:

Maximize $P = 2x + 1.5y$ subject to

$$3x + 4y \leq 1000$$

$$6x + 3y \leq 1200$$

$$x \geq 0, \quad y \geq 180$$

The graph of the feasible set S is shown in the figure that follows.

Vertex	$P = 2x + 1.5y$
$A(0,180)$	270
$B(0,250)$	375
$C(93\frac{1}{3},180)$	$456\frac{2}{3}$

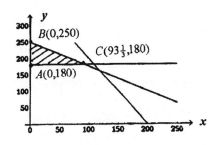

Thus, the optimal profit of $456 is realized when 93 units of model A hibachis and 180 units of model B hibachis are produced.

CHAPTER 4

EXERCISES 4.1, page 238

1. All entries in the last row of the simplex tableau are nonnegative and an optimal solution has been reached. We find
    $$x = 30/7, \ y = 20/7, \ u = 0, \ v = 0, \ \text{and} \ P = 220/7.$$

3. The simplex tableau is not in final form because there is an entry in the last row that is negative. The entry in the first row, second column, is the next pivot element and has a value of 1/2.

5. The simplex tableau is in final form. We find
    $$x = 1/3, \ y = 0, \ z = 13/3, \ u = 0, \ v = 6, \ w = 0 \ \text{and} \ P = 17.$$

7. The simplex tableau is not in final form because there are two entries in the last row that are negative. The entry in the third row, second column, is the pivot element and has a value of 1.

9. The simplex tableau is in final form. The solutions are
    $$x = 30, \ y = 0, \ z = 0, \ u = 10, \ v = 0, \ \text{and} \ P = 60,$$
    $$\text{and} \quad x = 0, \ y = 30, \ z = 0, \ u = 10, \ v = 0, \ \text{and} \ P = 60.$$
    (There are infinitely many answers).

11.

	x	y	u	v	P	Const	Ratio
p.r.→	1	②	1	0	0	12	$12/2 = 6$
	3	2	0	1	0	24	$24/2 = 12$
	−10	−12	0	0	1	0	

$$\underset{\text{p.c.}}{\uparrow}$$

$$\xrightarrow{\frac{1}{2}R_1}$$

	x	y	u	v	P	Const
	$\frac{1}{2}$	1	$\frac{1}{2}$	0	0	6
	3	2	0	1	0	24
	−10	−12	0	0	1	0

$$\xrightarrow[R_3 + 12R_1]{R_2 - 2R_1}$$

x	y	u	v	P	Const		Ratio
$\frac{1}{2}$	1	$\frac{1}{2}$	0	0	6		$6/(1/2)=12$
$p.r.\rightarrow$ ②	0	−1	1	0	12		$12/2=6$
−4	0	6	0	1	72		

$\uparrow$
$p.c.$

$\xrightarrow{\frac{1}{2}R_2}$

x	y	u	v	P	Const
$\frac{1}{2}$	1	$\frac{1}{2}$	0	0	6
1	0	$-\frac{1}{2}$	$\frac{1}{2}$	0	6
−4	0	6	0	1	72

$\xrightarrow[R_3+4R_2]{R_1-\frac{1}{2}R_2}$

x	y	u	v	P	Const
0	1	$\frac{3}{4}$	$-\frac{1}{4}$	0	3
1	0	$-\frac{1}{2}$	$\frac{1}{2}$	0	6
0	0	4	2	1	96

The last tableau is in final form. We find that $x = 6$, $y = 3$, $u = 0$, $v = 0$, and $P = 96$.

13. We obtain the following sequence of tableaus:

x	y	u	v	w	P	Const.		Ratio
3	1	1	0	0	0	24		24
2	1	0	1	0	0	18		18
$p.r.\rightarrow$ 1	③	0	0	1	0	24		8
−4	−6	0	0	0	1	0		

$\uparrow$
$p.c.$

$\xrightarrow{\frac{1}{3}R_3}$

x	y	u	v	w	P	Const.
3	1	1	0	0	0	24
2	1	0	1	0	0	18
$\frac{1}{3}$	1	0	0	$\frac{1}{3}$	0	8
−4	−6	0	0	0	1	0

$\xrightarrow[R_4+6R_3]{\substack{R_1-R_3 \\ R_2-R_3}}$

	x	y	u	v	w	P	Const.		Ratio
p.r.→	$\frac{8}{3}$	0	1	0	$-\frac{1}{3}$	0	16		6
	$\frac{5}{3}$	0	0	1	$-\frac{1}{3}$	0	10		6
	$\frac{1}{3}$	1	0	0	$\frac{1}{3}$	0	8		24
	-2	0	0	0	2	1	48		

↑
p.c.

(Observe that we have a choice here.)

$\xrightarrow{\frac{3}{8}R_1}$

x	y	u	v	w	P	Const.
1	0	$\frac{3}{8}$	0	$-\frac{1}{8}$	0	6
$\frac{5}{3}$	0	0	1	$-\frac{1}{3}$	0	10
$\frac{1}{3}$	1	0	0	$\frac{1}{3}$	0	8
-2	0	0	0	2	1	48

$\begin{array}{l} R_2-\frac{5}{3}R_1 \\ R_3-\frac{1}{3}R_1 \\ R_4+2R_1 \end{array} \longrightarrow$

x	y	u	v	w	P	Const.
1	0	$\frac{3}{8}$	0	$-\frac{1}{8}$	0	6
0	0	$-\frac{5}{8}$	1	$-\frac{1}{8}$	0	0
0	1	$-\frac{1}{8}$	0	$\frac{3}{8}$	0	6
0	0	$\frac{3}{4}$	0	$\frac{7}{4}$	1	60

We deduce that $x = 6$, $y = 6$, $u = 0$, $v = 0$, $w = 0$, and $P = 60$.

15.   We obtain the following sequence of tableaus:

	x	y	z	u	v	P	Const.		Ratio	
	1	1	1	1	0	0	8		$8/1 = 8$	$\frac{1}{4}R_2$
p.r.→	3	2	④	0	1	0	24		$24/4 = 6$	→
	-3	-4	-5	0	0	1	0			

↑
. p.c

x	y	z	u	v	P	Const.
1	1	1	1	0	0	8
$\frac{3}{4}$	$\frac{1}{2}$	1	0	$\frac{1}{4}$	0	6
-3	-4	-5	0	0	1	0

$\begin{array}{c} R_1 - R_2 \\ R_3 + 5R_2 \end{array} \longrightarrow$

	x	y	z	u	v	P	Const.		Ratio
$p.r. \rightarrow$	$\frac{1}{4}$	$\boxed{\frac{1}{2}}$	0	1	$-\frac{1}{4}$	0	2		$2/(1/2) = 4$
	$\frac{3}{4}$	$\frac{1}{2}$	1	0	$\frac{1}{4}$	0	6		$6/(1/2) = 12$
	$\frac{3}{4}$	$-\frac{3}{2}$	0	0	$\frac{5}{4}$	1	30		

$\underset{p.c.}{\uparrow}$ $\qquad \xrightarrow{2R_1}$

x	y	z	u	v	P	Const.
$\frac{1}{2}$	1	0	2	$-\frac{1}{2}$	0	4
$\frac{3}{4}$	$\frac{1}{2}$	1	0	$\frac{1}{4}$	0	6
$\frac{3}{4}$	$-\frac{3}{2}$	0	0	$\frac{5}{4}$	1	30

$\begin{array}{c} R_2 - \frac{1}{2}R_1 \\ R_3 + \frac{3}{2}R_1 \end{array} \longrightarrow$

x	y	z	u	v	P	Const.
$\frac{1}{2}$	1	0	2	$-\frac{1}{2}$	0	4
$\frac{1}{2}$	0	1	-1	$\frac{1}{2}$	0	4
$\frac{3}{2}$	0	0	3	$\frac{1}{2}$	1	36

This last tableau is in final form. We find that $x = 0$, $y = 4$, $z = 4$, $u = 0$, $v = 0$, and $P = 36$.

17.

	x	y	z	u	v	w	P	Const.		Ratio
	3	10	5	1	0	0	0	120		$120/10 = 12$
$p.r. \rightarrow$	5	$\boxed{2}$	8	0	1	0	0	6		$6/2 = 3$
	8	10	3	0	0	1	0	105		$105/10 = 21/2$
	-3	-4	-1	0	0	0	1	0		

$\qquad \underset{p.c.}{\uparrow}$ $\qquad \xrightarrow{\frac{1}{2}R_2}$

x	y	z	u	v	w	P	Const.
3	10	5	1	0	0	0	120
$\frac{5}{2}$	1	4	0	$\frac{1}{2}$	0	0	3
8	10	3	0	0	1	0	105
−3	−4	−1	0	0	0	1	0

$$\xrightarrow{\begin{array}{l} R_1-10R_2 \\ R_3-10R_2 \\ R_4+4R_2 \end{array}}$$

x	y	z	u	v	w	P	Const.
−22	0	−35	1	−5	0	0	90
$\frac{5}{2}$	1	4	0	$\frac{1}{2}$	0	0	3
−17	0	−37	0	−5	1	0	75
7	0	15	0	2	0	1	12

The last tableau is in final form. We find that $x = 0$, $y = 3$, $z = 0$, $u = 90$, $v = 0$, $w = 75$, and $P = 12$.

19. We obtain the following sequence of tableaus:

	x	y	z	u	v	w	P	Const.	Ratio
p.r.→	1	1	1	1	0	0	0	20	20
	2	④	3	0	1	0	0	42	$10\frac{1}{2}$
	2	0	3	0	0	1	0	30	− − −
	−4	−6	−5	0	0	0	1	0	

p.c. (below y column)

$$\xrightarrow{\frac{1}{4}R_2}$$

x	y	z	u	v	w	P	Const.
1	1	1	1	0	0	0	20
$\frac{1}{2}$	1	$\frac{3}{4}$	0	$\frac{1}{4}$	0	0	$\frac{21}{2}$
2	0	3	0	0	1	0	30
−4	−6	−5	0	0	0	1	0

$$\xrightarrow{\begin{array}{l} R_1-R_2 \\ R_4+6R_2 \end{array}}$$

	x	y	z	u	v	w	P	Const.		Ratio
	$\frac{1}{2}$	0	$\frac{1}{4}$	1	$-\frac{1}{4}$	0	0	$\frac{19}{2}$		19
	$\frac{1}{2}$	1	$\frac{3}{4}$	0	$\frac{1}{4}$	0	0	$\frac{21}{2}$		21
$p.r. \to$	②	0	3	0	0	1	0	30		15
	-1	0	$-\frac{1}{2}$	0	$\frac{3}{2}$	0	1	63		
	$\uparrow$									
	$p.c.$									

$\xrightarrow{\frac{1}{2}R_3}$

x	y	z	u	v	w	P	Const.
$\frac{1}{2}$	0	$\frac{1}{4}$	1	$-\frac{1}{4}$	0	0	$\frac{19}{2}$
$\frac{1}{2}$	1	$\frac{3}{4}$	0	$\frac{1}{4}$	0	0	$\frac{21}{2}$
1	0	$\frac{3}{2}$	0	0	$\frac{1}{2}$	0	15
-1	0	$-\frac{1}{2}$	0	$\frac{3}{2}$	0	1	63

$\xrightarrow[\substack{R_2-\frac{1}{2}R_3 \\ R_4+R_3}]{R_1-\frac{1}{2}R_3}$

x	y	z	u	v	w	P	Const.
0	0	$-\frac{1}{2}$	1	$-\frac{1}{4}$	$-\frac{1}{4}$	0	2
0	1	0	0	$\frac{1}{4}$	$-\frac{1}{4}$	0	3
1	0	$\frac{3}{2}$	0	0	$\frac{1}{2}$	0	15
0	0	1	0	$\frac{3}{2}$	$\frac{1}{2}$	1	78

So the solution is $x = 15$, $y = 3$, $z = 0$, $u = 2$, $v = 0$, $w = 0$, and $P = 78$.

21. We obtain the following sequence of tableaus:

	x	y	z	u	v	w	P	Const.		Ratio
$p.r. \to$	②	1	1	1	0	0	0	10		$10 / 2 = 5$
	3	5	1	0	1	0	0	45		$45 / 3 = 15$
	2	5	1	0	0	1	0	40		$40 / 2 = 20$
	-12	-10	-5	0	0	0	1	0		
	$\uparrow$									
	$p.c.$									

$\xrightarrow{\frac{1}{2}R_1}$

x	y	z	u	v	w	P	Const.
1	$\frac{1}{2}$	$\frac{1}{2}$	$\frac{1}{2}$	0	0	0	5
3	5	1	0	1	0	0	45
2	5	1	0	0	1	0	40
-12	-10	-5	0	0	0	1	0

$$\begin{array}{l} R_2-3R_1 \\ R_3-2R_1 \\ R_4+12R_1 \end{array} \longrightarrow$$

x	y	z	u	v	w	P	Const.		Ratio
1	$\frac{1}{2}$	$\frac{1}{2}$	$\frac{1}{2}$	0	0	0	5		$5/(1/2)=10$
0	$\frac{7}{2}$	$-\frac{1}{2}$	$-\frac{3}{2}$	1	0	0	30		$30/(7/2)=60/7$
0	④	0	-1	0	1	0	30		$30/4=15/2$
0	-4	1	6	0	0	1	60		

p.r.→ (row 3), p.c (column y)

$$\frac{1}{4}R_3 \longrightarrow$$

x	y	z	u	v	w	P	Const.
1	$\frac{1}{2}$	$\frac{1}{2}$	$\frac{1}{2}$	0	0	0	5
0	$\frac{7}{2}$	$-\frac{1}{2}$	$-\frac{3}{2}$	1	0	0	30
0	1	0	$-\frac{1}{4}$	0	$\frac{1}{4}$	0	$\frac{15}{2}$
0	-4	1	6	0	0	1	60

$$\begin{array}{l} R_1-\frac{1}{2}R_3 \\ R_2-\frac{7}{2}R_3 \\ R_4+4R_3 \end{array} \longrightarrow$$

x	y	z	u	v	w	P	Const.
1	0	$\frac{1}{2}$	$\frac{5}{8}$	0	$-\frac{1}{8}$	0	$\frac{5}{4}$
0	0	$-\frac{1}{2}$	$-\frac{5}{8}$	1	$-\frac{7}{8}$	0	$\frac{15}{4}$
0	1	0	$-\frac{1}{4}$	0	$\frac{1}{4}$	0	$\frac{15}{2}$
0	0	1	5	0	1	1	90

This last tableau is in final form, and we conclude that $x = 5/4$, $y = 15/2$, $z = 0$, $u = 0$, $v = 15/4$, $w = 0$, and $P = 90$.

23. We obtain the following sequence of tableaus, where u, v and w are slack variables.

	x	y	z	u	v	w	P	Const.		Ratio
p.r. →	(2)	1	2	1	0	0	0	7		$\frac{7}{2}$
	2	3	1	0	1	0	0	8		4
	1	2	3	0	0	1	0	7		7
	−24	−16	−23	0	0	0	1	0		

p.c. (under x) $\xrightarrow{\frac{1}{2}R_1}$

x	y	z	u	v	w	P	Const.
1	$\frac{1}{2}$	1	$\frac{1}{2}$	0	0	0	$\frac{7}{2}$
2	3	1	0	1	0	0	8
1	2	3	0	0	1	0	7
−24	−16	−23	0	0	0	1	0

$\xrightarrow[]{\begin{array}{c} R_2-2R_1 \\ R_3-R_1 \\ R_4+24R_1 \end{array}}$

	x	y	z	u	v	w	P	Const.		Ratio
	1	$\frac{1}{2}$	1	$\frac{1}{2}$	0	0	0	$\frac{7}{2}$		7
p.r. →	0	(2)	−1	−1	1	0	0	1		$\frac{1}{2}$
	0	$\frac{3}{2}$	2	$-\frac{1}{2}$	0	1	0	$\frac{7}{2}$		$\frac{7}{3}$
	0	−4	1	12	0	0	1	84		

p.c. (under y) $\xrightarrow{\frac{1}{2}R_2}$

x	y	z	u	v	w	P	Const.
1	$\frac{1}{2}$	1	$\frac{1}{2}$	0	0	0	$\frac{7}{2}$
0	1	$-\frac{1}{2}$	$-\frac{1}{2}$	$\frac{1}{2}$	0	0	$\frac{1}{2}$
0	$\frac{3}{2}$	2	$-\frac{1}{2}$	0	1	0	$\frac{7}{2}$
0	−4	1	12	0	0	1	84

$\xrightarrow[]{\begin{array}{c} R_1-\frac{1}{2}R_2 \\ R_3-\frac{3}{2}R_2 \\ R_4+4R_2 \end{array}}$

	x	y	z	u	v	w	P	Const.		Ratio
	1	0	$\frac{5}{4}$	$\frac{3}{4}$	$\frac{1}{4}$	0	0	$\frac{13}{4}$		$\frac{13}{5}$
	0	1	$-\frac{1}{2}$	$-\frac{1}{2}$	$\frac{1}{2}$	0	0	$\frac{1}{2}$		− −
p.r. →	0	0	$\left(\frac{11}{4}\right)$	$\frac{1}{4}$	$-\frac{3}{4}$	1	0	$\frac{11}{4}$		1
	0	0	−1	10	2	0	1	86		

p.c. (under z) $\xrightarrow{\frac{4}{11}R_3}$

x	y	z	u	v	w	P	Const.
1	0	$\frac{5}{4}$	$\frac{3}{4}$	$\frac{1}{4}$	0	0	$\frac{13}{4}$
0	1	$-\frac{1}{2}$	$-\frac{1}{2}$	$\frac{1}{2}$	0	0	$\frac{1}{2}$
0	0	1	$\frac{1}{11}$	$-\frac{3}{11}$	$\frac{4}{11}$	0	1
0	0	-1	10	2	0	1	86

$$\begin{array}{l} R_1-\frac{5}{4}R_3 \\ R_2+\frac{1}{2}R_3 \\ R_4+R_3 \end{array} \longrightarrow$$

x	y	z	u	v	w	P	Const.
1	0	0	$\frac{7}{11}$	$\frac{13}{22}$	$-\frac{5}{11}$	0	2
0	1	0	$-\frac{5}{11}$	$\frac{4}{11}$	$\frac{2}{11}$	0	1
0	0	1	$\frac{1}{11}$	$-\frac{3}{11}$	$\frac{4}{11}$	0	1
0	0	0	$\frac{111}{11}$	$\frac{19}{11}$	$\frac{4}{11}$	1	87

This last tableau is in final form and we conclude that P attains a maximum value of 87 when $x = 2$, $y = 1$, and $z = 1$.

25. Pivoting about the x-column in the initial simplex tableau, we have

	x	y	z	u	v	P	Const.		Ratio	
	3	3	-2	1	0	0	100		$100/3$	$\frac{1}{5}R_2$
p.r. $\rightarrow$	⑤	5	3	0	1	0	150		$150/5$	$\longrightarrow$
	-2	-2	4	0	0	1	0			

$$\uparrow$$
$$p.c.$$

x	y	z	u	v	P	Const.
3	3	-2	1	0	0	100
1	1	$\frac{3}{5}$	0	$\frac{1}{5}$	0	30
-2	-2	4	0	0	1	0

$$\begin{array}{l} R_1-3R_2 \\ R_3+2R_2 \end{array} \longrightarrow$$

x	y	z	u	v	P	Const.
0	0	$-\frac{19}{5}$	1	$-\frac{3}{5}$	0	10
1	1	$\frac{3}{5}$	0	$\frac{1}{5}$	0	30
0	0	$\frac{26}{5}$	0	$\frac{2}{5}$	1	60

and we see that one optimal solution occurs when $x = 30$, $y = 0$, $z = 0$, and $P = 60$. Similarly, pivoting about the y-column, we obtain another optimal solution: $x = 0$, $y = 30$, $z = 0$, and $P = 60$.

27. Let the number of model A and model B fax machines made each month be x and y, respectively. Then we have the following linear programming problem:

$$\text{Maximize } P = 25x + 40y \text{ subject to}$$
$$200x + 300y \le 600{,}000$$
$$x + y \le 2{,}500$$
$$x \ge 0,\ y \ge 0$$

Using the simplex method, we obtain the following sequence of tableaus:

x	y	u	v	P	Const		Ratio	
200	⓷⓪⓪	1	0	0	600,000		2000	$\frac{1}{300}R_1$
1	1	0	1	0	2,500		2500	
−25	−40	0	0	1	0			

x	y	u	v	P	Const	
$\frac{2}{3}$	1	$\frac{1}{300}$	0	0	2000	$\begin{array}{c} R_2 - R_1 \\ R_3 + 40R_1 \end{array}$
1	1	0	1	0	2500	
−25	−40	0	0	1	0	

x	y	u	v	P	Const
$\frac{2}{3}$	1	$\frac{1}{300}$	0	0	2000
$\frac{1}{3}$	0	$-\frac{1}{300}$	1	0	500
$\frac{5}{3}$	0	$\frac{2}{15}$	0	1	80,000

We conclude that the maximum monthly profit is \$80,000, and this occurs when 0 model A and 2000 model B fax machines are produced.

29. Suppose the farmer plants x acres of Crop A and y acres of crop B. Then the problem is

Maximize $P = 150x + 200y$ subject to
$$x + y \leq 150$$
$$40x + 60y \leq 7400$$
$$20x + 25y \leq 3300$$
$$x \geq 0,\ y \geq 0$$

Using the simplex method, we obtain the following sequence of tableaus:

x	y	u	v	w	P	Const	
①	1	1	0	0	0	150	$R_2 - 40R_1$
40	60	0	1	0	0	7400	$R_3 - 20R_1$
20	25	0	0	1	0	3300	$R_4 + 150R_1$
-150	-200	0	0	0	1	0	$\longrightarrow$

	x	y	u	v	w	P	Const	Ratio	
	1	1	1	0	0	0	150	150	
	0	20	-40	1	0	0	1400	700	$\frac{1}{5}R_3$
$p.r. \rightarrow$	0	⑤	-20	0	1	0	300	60	$\longrightarrow$
	0	-50	150	0	0	1	22500		

$\uparrow$
$p.c.$

x	y	u	v	w	P	Const	
1	1	1	0	0	0	150	$R_1 - R_3$
0	20	-40	1	0	0	1400	$R_2 - 20R_3$
0	1	-4	0	$\frac{1}{5}$	0	60	$R_4 + 50R_3$
0	-50	150	0	0	1	22500	$\longrightarrow$

	x	y	u	v	w	P	Const	Ratio
	1	0	5	0	$-\frac{1}{5}$	0	90	18
p.r. →	0	0	(40)	1	-4	0	200	5
	0	1	-4	0	$\frac{1}{5}$	0	60	$--$
	0	0	-50	0	10	1	25500	

$\xrightarrow{\frac{1}{40}R_2}$

p.c.

x	y	u	v	w	P	Const
1	0	5	0	$-\frac{1}{5}$	0	90
0	0	1	$\frac{1}{40}$	$-\frac{1}{10}$	0	5
0	1	-4	0	$\frac{1}{5}$	0	60
0	0	-50	0	10	1	25500

$\begin{array}{c} R_1-5R_2 \\ R_3+4R_2 \\ R_4+50R_2 \end{array} \longrightarrow$

x	y	u	v	w	P	Const
1	0	0	$-\frac{1}{8}$	$\frac{3}{10}$	0	65
0	0	1	$\frac{1}{40}$	$-\frac{1}{10}$	0	5
0	1	0	$\frac{1}{10}$	$-\frac{1}{5}$	0	80
0	0	0	$\frac{5}{4}$	5	1	25750

The last tableau is in final form. We find $x = 65$, $y = 80$, and $P = 25,750$. So the maximum profit of \$25,750 is realized by planting 65 acres of Crop A and 80 acres of Crop B.

31. Suppose the Excelsior Company buys x, y, and z minutes of morning, afternoon, and evening commercials, respectively. Then we wish to maximize
$$P = 200,000x + 100,000y + 600,000z \text{ subject to}$$
$$3000x + 1000y + 12,000z \leq 102,000$$
$$z \leq 6$$
$$x + y + z \leq 25$$
$$x \geq 0, y \geq 0, z \geq 0$$

Using the simplex method, we obtain the following sequence of tableaus.

x	y	z	u	v	w	P	Const.		Ratio	
3000	1000	12,000	1	0	0	0	102,000		17/2	$R_1 - 12{,}000R_2$
0	0	1	0	1	0	0	6		6	$R_3 - R_2$
1	1	1	0	0	1	0	25		25	$R_4 + 600{,}000R_2$
$-200{,}000$	$-100{,}000$	$-600{,}000$	0	0	0	1	0			$\longrightarrow$

x	y	z	u	v	w	P	Const.		Ratio	
⟨3000⟩	1000	0	1	$-12{,}000$	0	0	30,000		10	
0	0	1	0	1	0	0	6		— —	$\frac{1}{3000}R_1$
1	1	0	0	-1	1	0	19		19	$\longrightarrow$
$-200{,}000$	$-100{,}000$	0	0	600,000	0	1	3,600,000			

x	y	z	u	v	w	P	Const.	
1	$\frac{1}{3}$	0	$\frac{1}{3000}$	-4	0	0	10	
0	0	1	0	1	0	0	6	$R_3 - R_1$
1	1	0	0	-1	1	0	19	$R_4 + 200{,}000R_1$
$-200{,}000$	$-100{,}000$	0	0	600,000	0	1	3,600,000	$\longrightarrow$

x	y	z	u	v	w	P	Const.	
1	$\frac{1}{3}$	0	$\frac{1}{3000}$	-4	0	0	10	
0	0	1	0	1	0	0	6	$\frac{1}{3}R_3$
0	$\frac{2}{3}$	0	$-\frac{1}{3000}$	⟨3⟩	1	0	9	$\longrightarrow$
0	$-\frac{100{,}000}{3}$	0	$\frac{200}{3}$	$-200{,}000$	0	1	5,600,000	

x	y	z	u	v	w	P	Const.	
1	$\frac{1}{3}$	0	$\frac{1}{3000}$	-4	0	0	10	$R_1 + 4R_3$
0	0	1	0	1	0	0	6	$R_2 - R_3$
0	$\frac{2}{9}$	0	$-\frac{1}{9000}$	1	$\frac{1}{3}$	0	3	$R_4 + 200{,}000R_3$
0	$-\frac{100{,}000}{3}$	0	$\frac{200}{3}$	$-200{,}000$	0	1	5,600,000	$\longrightarrow$

x	y	z	u	v	w	P	Const.
1	$\frac{11}{9}$	0	$-\frac{1}{9000}$	0	$\frac{4}{3}$	0	22
0	$-\frac{2}{9}$	1	$\frac{1}{9000}$	0	$-\frac{1}{3}$	0	3
0	$\frac{2}{9}$	0	$-\frac{1}{9000}$	1	$\frac{1}{3}$	0	3
0	$\frac{100,000}{9}$	0	$\frac{400}{9}$	0	$\frac{200,000}{3}$	1	6,200,000

We conclude that $x = 22$, $y = 0$, $z = 3$, $u = 0$, $v = 3$, and $P = 6,200,000$. Therefore, the company should buy 22 minutes of morning and 3 minutes of evening advertising time, thereby maximizing their exposure to 6,200,000 viewers.

33. We first tabulate the given information:

MODELS

Dept.	A	B	C	Time available
Fabrication	$\frac{5}{4}$	$\frac{3}{2}$	$\frac{3}{2}$	310
Assembly	1	1	$\frac{3}{4}$	205
Finishing	1	1	$\frac{1}{2}$	190
Profit	26	28	24	

Let x, y, and z denote the number of units of model A, model B, and model C to be produced, respectively. Then the required linear programming problem is

Maximize $P = 26x + 28y + 24z$ subject to
$$\frac{5}{4}x + \frac{3}{2}y + \frac{3}{2}z \le 310$$
$$x + y + \frac{3}{4}z \le 205$$
$$x + y + \frac{1}{2}z \le 190$$
$$x \ge 0, y \ge 0, z \ge 0$$

Using the simplex method, we obtain the following tableaus:

x	y	z	u	v	w	P	Const.		Ratio	
$\frac{5}{4}$	$\frac{3}{2}$	$\frac{3}{2}$	1	0	0	0	310		$206\frac{2}{3}$	$R_1 - \frac{3}{2}R_3$
1	1	$\frac{3}{4}$	0	1	0	0	205		205	$R_2 - R_3$
$p.r. \rightarrow$ 1	①	$\frac{1}{2}$	0	0	1	0	190		190	$\xrightarrow{R_4 + 28R_3}$
-26	-28	-24	0	0	0	1	0			

$$\uparrow \atop p.c.$$

x	y	z	u	v	w	P	Const.		Ratio	
$p.r. \rightarrow$ $-\frac{1}{4}$	0	③⁄₄	1	0	$-\frac{3}{2}$	0	25		$33\frac{1}{3}$	
0	0	$\frac{1}{4}$	0	1	-1	0	15		60	$\xrightarrow{\frac{4}{3}R_1}$
1	1	$\frac{1}{2}$	0	0	1	0	190		380	
2	0	-10	0	0	28	1	5320			

$$\uparrow \atop p.c.$$

x	y	z	u	v	w	P	Const.	
$-\frac{1}{3}$	0	1	$\frac{4}{3}$	0	-2	0	$\frac{100}{3}$	$R_2 - \frac{1}{4}R_1$
0	0	$\frac{1}{4}$	0	1	-1	0	15	$R_3 - \frac{1}{2}R_1$
1	1	$\frac{1}{2}$	0	0	1	0	190	$\xrightarrow{R_4 + 10R_1}$
2	0	-10	0	0	28	1	5320	

x	y	z	u	v	w	P	Const.		Ratio	
$-\frac{1}{3}$	0	1	$\frac{4}{3}$	0	-2	0	$\frac{100}{3}$		$--$	
$p.r. \rightarrow$ ①⁄₁₂	0	0	$-\frac{1}{3}$	1	$-\frac{1}{2}$	0	$\frac{20}{3}$		80	$\xrightarrow{12R_2}$
$\frac{7}{6}$	1	0	$-\frac{2}{3}$	0	2	0	$\frac{520}{3}$		$148\frac{4}{7}$	
$-\frac{4}{3}$	0	0	$\frac{40}{3}$	0	8	1	$\frac{16960}{3}$			

$$\uparrow \atop p.c.$$

x	y	z	u	v	w	P	Const.	
$-\frac{1}{3}$	0	1	$\frac{4}{3}$	0	-2	0	$\frac{100}{3}$	$R_1 + \frac{1}{3}R_2$
1	0	0	-4	12	-6	0	80	$R_3 - \frac{7}{6}R_2$
$\frac{7}{6}$	1	0	$-\frac{2}{3}$	0	2	0	$\frac{520}{3}$	$\xrightarrow{R_4 + \frac{4}{3}R_2}$
$-\frac{4}{3}$	0	0	$\frac{40}{3}$	0	8	1	$\frac{16960}{3}$	

x	y	z	u	v	w	P	Const.
0	0	1	0	4	−4	0	60
1	0	0	−4	12	−6	0	80
0	1	0	4	−14	9	0	80
0	0	0	8	16	0	1	5760

The last tableau is in final form. We see that $x = 80$, $y = 80$, $z = 60$, and $P = 5760$. So, by producing 80 units each of Models A and B, and 60 units of Model C, the company stands to make a profit of $5760.

35. Let x, y, and z denote the number (in thousands) of bottles of formula I, formula II, and formula III, respectively, produced. The resulting linear programming problem is

$$\text{Maximize } P = 180x + 200y + 300z \text{ subject to}$$
$$\tfrac{5}{2}x + 3y + 4z \le 70$$
$$x \le 9$$
$$y \le 12$$
$$z \le 6$$
$$x \ge 0,\ y \ge 0,\ z \ge 0$$

Using the simplex method, we have

	x	y	z	s	t	u	v	P	Const.	Ratio	
	$\frac{5}{2}$	3	4	1	0	0	0	0	70	$17\frac{1}{2}$	
	1	0	0	0	1	0	0	0	9	--	$R_1 - 4R_4$
	0	1	0	0	0	1	0	0	12	--	$R_5 + 300R_4$
p.r.→	0	0	①	0	0	0	1	0	6	6	
	−180	−200	−300	0	0	0	0	1	0		

p.c.

	x	y	z	s	t	u	v	P	Const.	Ratio	
	$\frac{5}{2}$	3	0	1	0	0	-4	0	46	$15\frac{1}{3}$	
	1	0	0	0	1	0	0	0	9	--	R_1-3R_3
p.r.→	0	①	0	0	0	1	0	0	12	12	R_5+200R_3 →
	0	0	1	0	0	0	1	0	6	--	
	-180	-200	0	0	0	0	300	1	1800		

p.c. ↑

x	y	z	s	t	u	v	P	Const.	Ratio	
⑤⁄₂	0	0	1	0	-3	-4	0	10	4	
1	0	0	0	1	0	0	0	9	9	$\frac{2}{5}R_1$ →
0	1	0	0	0	1	0	0	12	--	
0	0	1	0	0	0	1	0	6	--	
-180	0	0	0	0	200	300	1	4200		

x	y	z	s	t	u	v	P	Const.	
1	0	0	$\frac{2}{5}$	0	$-\frac{6}{5}$	$-\frac{8}{5}$	0	4	
1	0	0	0	1	0	0	0	9	R_2-R_1
0	1	0	0	0	1	0	0	12	R_5+180R_1 →
0	0	1	0	0	0	1	0	6	
-180	0	0	0	0	200	300	1	4200	

x	y	z	s	t	u	v	P	Const.	Ratio	
1	0	0	$\frac{2}{5}$	0	$-\frac{6}{5}$	$-\frac{8}{5}$	0	4	--	
1	0	0	$-\frac{2}{5}$	1	⑥⁄₅	$\frac{8}{5}$	0	5	$\frac{25}{6}$	$\frac{5}{6}R_2$ →
0	1	0	0	0	1	0	0	12	12	
0	0	1	0	0	0	1	0	6	--	
0	0	0	72	0	-16	12	1	4920		

x	y	z	s	t	u	v	P	Const.
1	0	0	$\frac{2}{5}$	0	$-\frac{6}{5}$	$-\frac{8}{5}$	0	4
0	0	0	$-\frac{1}{3}$	$\frac{5}{6}$	1	$\frac{4}{3}$	0	$\frac{25}{6}$
0	1	0	0	0	1	0	0	12
0	0	1	0	0	0	1	0	6
0	0	0	72	0	-16	12	1	4920

$$R_1 + \frac{6}{5}R_2$$
$$R_3 - R_2$$
$$R_5 + 16R_2 \longrightarrow$$

x	y	z	s	t	u	v	P	Const.
1	0	0	0	1	0	0	0	9
0	0	0	$-\frac{1}{3}$	$\frac{5}{6}$	1	$\frac{4}{3}$	0	$\frac{25}{6}$
0	1	0	$\frac{1}{3}$	$-\frac{5}{6}$	0	$-\frac{4}{3}$	0	$\frac{47}{6}$
0	0	1	0	0	0	1	0	6
0	0	0	$\frac{200}{3}$	$\frac{40}{3}$	0	$\frac{100}{3}$	1	$4986\frac{2}{3}$

Therefore, $x = 9$, $y = 47/6$, $z = 6$, and $P \approx 4986.67$; that is, the company should manufacture 9000 bottles of formula *I*, 7833 bottles of formula *II*, and 6000 bottles of formula *III* for a maximum profit of $4986.67.

37. True. See the explanation of the simplex method on page 226.

USING TECHNOLOGY EXERCISES 4.1, page 243

1. $x = 1.2$, $y = 0$, $z = 1.6$, $w = 0$, and $P = 8.8$

3. $x = 1.6$, $y = 0$, $z = 0$, $w = 3.6$, and $P = 12.4$

EXERCISES 4.2, page 256

1. We solve the associated regular problem:
    Maximize $P = -C = 2x - y$ subject to
    $$x + 2y \leq 6$$
    $$3x + 2y \leq 12$$
    $$x \geq 0, \ y \geq 0$$
    Using the simplex method where u and v are slack variables, we have

	x	y	u	v	P	Const.	Ratio	
	1	2	1	0	0	6	6	$\frac{1}{3}R_2$
p.r. →	③	2	0	1	0	12	4	→
	-2	1	0	0	1	0		

↑
p.c.

x	y	u	v	P	Const.	
1	2	1	0	0	6	$R_1 - R_2$
1	$\frac{2}{3}$	0	$\frac{1}{3}$	0	4	$R_3 + 2R_2$ →
-2	1	0	0	1	0	

x	y	u	v	P	Const.
0	$\frac{4}{3}$	1	$-\frac{1}{3}$	0	2
1	$\frac{2}{3}$	0	$\frac{1}{3}$	0	4
0	$\frac{7}{3}$	0	$\frac{2}{3}$	1	8

Therefore, $x = 4$, $y = 0$, and $C = -P = -8$.

3. We maximize $P = -C = 3x + 2y$. Using the simplex method, we obtain

	x	y	u	v	P	Const.	Ratio	
	3	4	1	0	0	24	8	$\frac{1}{7}R_2$
p.r.→	⑦	-4	0	1	0	16	$\frac{16}{7}$	→
	-3	-2	0	0	1	0		

↑
p.c.

x	y	u	v	P	Const.	
3	4	1	0	0	24	$R_1 - 3R_2$
1	$-\frac{4}{7}$	0	$\frac{1}{7}$	0	$\frac{16}{7}$	$R_3 + 3R_2$ →
-3	-2	0	0	1	0	

	x	y	u	v	P	Const.	Ratio	
p.r.→	0	(40/7)	1	$-\frac{3}{7}$	0	$\frac{120}{7}$	3	$\frac{7}{40}R_1$ →
	1	$-\frac{4}{7}$	0	$\frac{1}{7}$	0	$\frac{16}{7}$	--	
	0	$-\frac{26}{7}$	0	$\frac{3}{7}$	1	$\frac{48}{7}$		

p.c.

x	y	u	v	P	Const.	
0	1	$\frac{7}{40}$	$-\frac{3}{40}$	0	3	$R_2+\frac{4}{7}R_1$
1	$-\frac{4}{7}$	0	$\frac{1}{7}$	0	$\frac{16}{7}$	$R_3+\frac{26}{7}R_1$ →
0	$-\frac{26}{7}$	0	$\frac{3}{7}$	1	$\frac{48}{7}$	

x	y	u	v	P	Const.
0	1	$\frac{7}{40}$	$-\frac{3}{40}$	0	3
1	0	$\frac{1}{10}$	$\frac{1}{10}$	0	4
0	0	$\frac{13}{20}$	$\frac{3}{20}$	1	18

The last tableau is in final form. We find $x = 4$, $y = 3$, and $C = -P = -18$.

5. We maximize $P = -C = -2x + 3y + 4z$ subject to the given constraints. Using the simplex method we obtain

	x	y	z	u	v	w	P	Const.	Ratio	
	-1	2	-1	1	0	0	0	8	--	
p.r. →	1	-2	(2)	0	1	0	0	10	5	$\frac{1}{2}R_2$ →
	2	4	-3	0	0	1	0	12	--	
	2	-3	-4	0	0	0	1	0		

p.c.

x	y	z	u	v	w	P	Const.	
-1	2	-1	1	0	0	0	8	R_1+R_2
$\frac{1}{2}$	-1	1	0	$\frac{1}{2}$	0	0	5	R_3+3R_2
2	4	-3	0	0	1	0	12	R_4+4R_2 →
2	-3	-4	0	0	0	1	0	

	x	y	z	u	v	w	P	Const.	Ratio	
$p.r. \rightarrow$	$-\frac{1}{2}$	①	0	1	$\frac{1}{2}$	0	0	13	13	R_2+R_1
	$\frac{1}{2}$	-1	1	0	$\frac{1}{2}$	0	0	5	$--$	R_3-R_1
	$\frac{7}{2}$	1	0	0	$\frac{3}{2}$	1	0	27	27	$R_4+7R_1 \rightarrow$
	4	-7	0	0	2	0	1	20		

$\uparrow$ $p.c.$

x	y	z	u	v	w	P	Const.
$-\frac{1}{2}$	1	0	1	$\frac{1}{2}$	0	0	13
0	0	1	1	1	0	0	18
4	0	0	-1	1	1	0	14
$\frac{1}{2}$	0	0	7	$\frac{11}{2}$	0	1	111

The last tableau is in final form. We see that $x = 0$, $y = 13$, $z = 18$, $w = 14$, and $C = -P = -111$.

7. $x = 5/4$, $y = 1/4$, $u = 2$, $v = 3$, and $C = P = 13$.

9. $x = 5$, $y = 10$, $z = 0$, $u = 1$, $v = 2$, and $C = P = 80$.

11. We first write the tableau

x	y	Const.
1	2	4
3	2	6
2	5	

Then obtain the following by interchanging rows and columns:

u	v	Const.
1	3	2
2	2	5
4	6	

From this table we construct the dual problem:

Maximize the objective function
$$P = 4u + 6v \text{ subject to}$$
$$u + 3v \le 2$$
$$2u + 2v \le 5$$
$$u \ge 0, \ v \ge 0$$

Solving the dual problem using the simplex method with x and y as the slack variables, we obtain

	u	v	x	y	P	Const.	Ratio	
p.r.$\rightarrow$	1	③	1	0	0	2	$\frac{2}{3}$	$\frac{1}{3}R_1$
	2	2	0	1	0	5	$\frac{5}{2}$	$\longrightarrow$
	−4	−6	0	0	1	0		

p.c. $\uparrow$

u	v	x	y	P	Const.	
$\frac{1}{3}$	1	$\frac{1}{3}$	0	0	$\frac{2}{3}$	$R_2 - 2R_1$
2	2	0	1	0	5	$R_3 + 6R_1$
−4	−6	0	0	1	0	$\longrightarrow$

	u	v	x	y	P	Const.	Ratio	
p.r.$\rightarrow$	$\left(\frac{1}{3}\right)$	1	$\frac{1}{3}$	0	0	$\frac{2}{3}$	2	$3R_1$
	$\frac{4}{3}$	0	$-\frac{2}{3}$	1	0	$\frac{11}{3}$	$2\frac{3}{4}$	$\longrightarrow$
	−2	0	2	0	1	4		

p.c. $\uparrow$

u	v	x	y	P	Const.	
1	3	1	0	0	2	$R_2 - \frac{4}{3}R_1$
$\frac{4}{3}$	0	$-\frac{2}{3}$	1	0	$\frac{11}{3}$	$R_3 + 2R_1$
−2	0	2	0	1	4	$\longrightarrow$

u	v	x	y	P	Const.
1	3	1	0	0	2
0	−4	−2	1	0	1
0	6	4	0	1	8

Interpreting the final tableau, we see that $x = 4$, $y = 0$, and $P = C = 8$.

13.   We first write the tableau

x	y	Const.
6	1	60
2	1	40
1	1	30
6	4	

Then obtain the following by interchanging rows and columns:

u	v	w	Const.
6	2	1	6
1	1	1	4
60	40	30	

From this table we construct the dual problem:
Maximize $P = 60u + 40v + 30w$ subject to
$$6u + 2v + w \le 6$$
$$u + v + w \le 4$$
$$u \ge 0, \ v \ge 0, \ w \ge 0$$

We solve the problem as follows.

	u	v	w	x	y	P	Const.	Ratio
p.r. →	⑥	2	1	1	0	0	6	1
	1	1	1	0	1	0	4	4
	−60	−40	−30	0	0	1	0	−−

$\frac{1}{6}R_1$

p.c ↑

u	v	w	x	y	P	Const.
1	$\frac{1}{3}$	$\frac{1}{6}$	$\frac{1}{6}$	0	0	1
1	1	1	0	1	0	4
−60	−40	−30	0	0	1	0

$R_2 - R_1$
$R_3 + 60R_1$

	u	v	w	x	y	P	Const.	Ratio	
	1	$\frac{1}{3}$	$\frac{1}{6}$	$\frac{1}{6}$	0	0	1	6	$\frac{6}{5}R_2$
p.r.→	0	$\frac{2}{3}$	$\left(\frac{5}{6}\right)$	$-\frac{1}{6}$	1	0	3	18/5	→
	0	−20	−20	10	0	1	60	− −	

p.c. (under w)

u	v	w	x	y	P	Const.	
1	$\frac{1}{3}$	$\frac{1}{6}$	$\frac{1}{6}$	0	0	1	$R_1 - \frac{1}{6}R_2$
0	$\frac{4}{5}$	1	$-\frac{1}{5}$	$\frac{6}{5}$	0	$\frac{18}{5}$	$R_3 + 20R_2$ →
0	−20	−20	10	0	1	60	

	u	v	w	x	y	P	Const.	Ratio	
p.r.→	1	$\left(\frac{1}{5}\right)$	0	$\frac{1}{5}$	$-\frac{1}{5}$	0	$\frac{2}{5}$	2	$5R_1$
	0	$\frac{4}{5}$	1	$-\frac{1}{5}$	$\frac{6}{5}$	0	$\frac{18}{5}$	9/2	→
	0	−4	0	6	24	1	132	− −	

p.c. (under v)

u	v	w	x	y	P	Const.	
5	1	0	1	−1	0	2	$R_2 - \frac{4}{5}R_1$
0	$\frac{4}{5}$	1	$-\frac{1}{5}$	$\frac{6}{5}$	0	$\frac{18}{5}$	$R_3 + 4R_1$ →
0	−4	0	6	24	1	132	

u	v	w	x	y	P	Const.
5	1	0	1	−1	0	2
−4	0	1	−1	2	0	2
20	0	0	10	20	1	140

The last tableau is in final form. We find that $x = 10$, $y = 20$, and $C = 140$.

15. We first write the tableau

x	y	z	Const.
20	10	1	10
1	1	2	20
200	150	120	

Then obtain the following by interchanging rows and columns:

u	v	Const.
20	1	200
10	1	150
1	2	120
10	20	

From this table we construct the dual problem:

Maximize $P = 10u + 20v$ subject to

$$20u + \ v \le 200$$
$$10u + \ v \le 150$$
$$u + 2v \le 120$$
$$u \ge 0,\ v \ge 0$$

Solving this problem, we obtain the following tableaus:

	u	v	x	y	z	P	Const.	Ratio	
	20	1	1	0	0	0	200	200	
	10	1	0	1	0	0	150	150	$\xrightarrow{\frac{1}{2}R_3}$
p.r.→	1	②	0	0	1	0	120	60	
	−10	−20	0	0	0	1	0		

$$\underset{p.\varsigma}{\uparrow}$$

u	v	x	y	z	P	Const.	
20	1	1	0	0	0	200	$\begin{array}{l}R_1-R_3\\ R_2-R_3\end{array}$
10	1	0	1	0	0	150	$\xrightarrow{R_4+20R_3}$
$\frac{1}{2}$	1	0	0	$\frac{1}{2}$	0	60	
−10	−20	0	0	0	1	0	

u	v	x	y	z	P	Const.
$\frac{39}{2}$	0	1	0	$-\frac{1}{2}$	0	140
$\frac{19}{2}$	0	0	1	$-\frac{1}{2}$	0	90
$\frac{1}{2}$	1	0	0	$\frac{1}{2}$	0	60
0	0	0	0	10	1	1200

This last tableau is in final form. We find that $x = 0$, $y = 0$, $z = 10$, and $C = 1200$.

17.   We first write the tableau

x	y	z	Const.
1	2	2	10
2	1	1	24
1	1	1	16
6	8	4	

Then obtain the following by interchanging rows and columns:

u	v	w	Const.
1	2	1	6
2	1	1	8
2	1	1	4
10	24	16	

From this table we construct the dual problem:
Maximize the objective function
$P = 10u + 24v + 16w$ subject to
$$u + 2v + w \le 6$$
$$2u + v + w \le 8$$
$$2u + v + w \le 4$$
$$u \ge 0, v \ge 0, w \ge 0$$

Solving the dual problem using the simplex method with x, y, and z as slack variables, we obtain

	u	v	w	x	y	z	P	Const.		Ratio
p.r. →	1	②	1	1	0	0	0	6		3
	2	1	1	0	1	0	0	8		8
	2	1	1	0	0	1	0	4		4
	−10	−24	−16	0	0	0	1	0		

$\xrightarrow{\frac{1}{2}R_1}$

↑
p.c.

u	v	w	x	y	z	P	Const.	
$\frac{1}{2}$	1	$\frac{1}{2}$	$\frac{1}{2}$	0	0	0	3	R_2-R_1
2	1	1	0	1	0	0	8	R_3-R_1 R_4+24R_1
2	1	1	0	0	1	0	4	
−10	−24	−16	0	0	0	1	0	

	u	v	w	x	y	z	P	Const.	Ratio	
	$\frac{1}{2}$	1	$\frac{1}{2}$	$\frac{1}{2}$	0	0	0	3	6	
	$\frac{3}{2}$	0	$\frac{1}{2}$	$-\frac{1}{2}$	1	0	0	5	10	$2R_3$
p.r.→	$\frac{3}{2}$	0	$\left(\frac{1}{2}\right)$	$-\frac{1}{2}$	0	1	0	1	2	
	2	0	−4	12	0	0	1	72		

$$\uparrow$$
$$p.c.$$

u	v	w	x	y	z	P	Const.	
$\frac{1}{2}$	1	$\frac{1}{2}$	$\frac{1}{2}$	0	0	0	3	$R_1-\frac{1}{2}R_3$
$\frac{3}{2}$	0	$\frac{1}{2}$	$-\frac{1}{2}$	1	0	0	5	$R_2-\frac{1}{2}R_3$
3	0	1	−1	0	2	0	2	R_4+4R_3
2	0	−4	12	0	0	1	72	

u	v	w	x	y	z	P	Const.
−1	1	0	1	0	−1	0	2
0	0	0	0	1	−1	0	4
3	0	1	−1	0	2	0	2
14	0	0	8	0	8	1	80

The solution to the primal problem is $x = 8$, $y = 0$, $z = 8$, and $C = 80$.

19. We first write

Maximize $P = 6u + 2v + 4w$ subject to the constraints

x	y	z	Const.
2	4	3	6
6	0	1	2
0	6	2	4
30	12	20	

Then obtain the following by interchanging rows and columns:

u	v	w	Const.
2	6	0	30
4	0	6	12
3	1	2	20
6	2	4	

From this table we construct the dual problem:

Maximize $P = 6u + 2v + 4w$ subject to

$$2u + 6v \qquad \leq 30$$
$$4u + \qquad 6w \leq 12$$
$$3u + v + 2w \leq 20$$
$$u \geq 0,\ v \geq 0,\ w \geq 0$$

Using the simplex method, we obtain

	u	v	w	x	y	z	P	Const.		Ratio
	2	6	0	1	0	0	0	30		15
p.r. →	④	0	6	0	1	0	0	12		3
	3	1	2	0	0	1	0	20		$\frac{20}{3}$
	−6	−2	−4	0	0	0	1	0		

$\frac{1}{4}R_2 \longrightarrow$

↑
p.c.

u	v	w	x	y	z	P	Const.
2	6	0	1	0	0	0	30
1	0	$\frac{3}{2}$	0	$\frac{1}{4}$	0	0	3
3	1	2	0	0	1	0	20
−6	−2	−4	0	0	0	1	0

$\begin{array}{c} R_1-2R_2 \\ R_3-3R_2 \\ R_4+6R_2 \end{array} \longrightarrow$

	u	v	w	x	y	z	P	Const.		Ratio
p.r.→	0	⑥	−3	1	$-\frac{1}{2}$	0	0	24		4
	1	0	$\frac{3}{2}$	0	$\frac{1}{4}$	0	0	3		---
	0	1	$-\frac{5}{2}$	0	$-\frac{3}{4}$	1	0	11		11
	0	−2	5	0	$\frac{3}{2}$	0	1	18		

$\begin{array}{c} \uparrow \\ \text{p.c.} \end{array}$ $\xrightarrow{\frac{1}{6}R_1}$

u	v	w	x	y	z	P	Const.
0	1	$-\frac{1}{2}$	$\frac{1}{6}$	$-\frac{1}{12}$	0	0	4
1	0	$\frac{3}{2}$	0	$\frac{1}{4}$	0	0	3
0	1	$-\frac{5}{2}$	0	$-\frac{3}{4}$	1	0	11
0	−2	5	0	$\frac{3}{2}$	0	1	18

$\begin{array}{c} R_3-R_1 \\ R_4+2R_1 \end{array} \longrightarrow$

u	v	w	x	y	z	P	Const.
0	1	$-\frac{1}{2}$	$\frac{1}{6}$	$-\frac{1}{12}$	0	0	4
1	0	$\frac{3}{2}$	0	$\frac{1}{4}$	0	0	3
0	0	−2	$-\frac{1}{6}$	$-\frac{2}{3}$	1	0	7
0	0	4	$\frac{1}{3}$	$\frac{4}{3}$	0	1	26

The last tableau is in final form. We find $x = 1/3$, $y = 4/3$, $z = 0$, and $C = 26$.

21. This problem was formulated in Exercise 12, Section 3.2, page 200 of the text.. We rewrite the constraints in the form

$$
\begin{aligned}
-x_1 - x_2 - x_3 & & & \geq -800 \\
& -x_4 - x_5 - x_6 & & \geq -600 \\
x_1 & + x_4 & & \geq 500 \\
x_2 & + x_5 & & \geq 400 \\
x_3 & + x_6 & & \geq 400
\end{aligned}
$$

We solve this problem using duality. We first write

x_1	x_2	x_3	x_4	x_5	x_6	Const.
-1	-1	-1	0	0	0	-800
0	0	0	-1	-1	-1	-600
1	0	0	1	0	0	500
0	1	0	0	1	0	400
0	0	1	0	0	1	400
16	20	22	18	16	14	

Interchanging the rows with the columns, we obtain

u_1	u_2	u_3	u_4	u_5	Const.
-1	0	1	0	0	16
-1	0	0	1	0	20
-1	0	0	0	1	22
0	-1	1	0	0	18
0	-1	0	1	0	16
0	-1	0	0	1	14
-800	-600	500	400	400	

from which we obtain the dual problem:

Maximize $P = -800u_1 - 600u_2 + 500u_3 + 400u_4 + 400u_5$ subject to

$$-u_1 \quad\; +u_3 \qquad\qquad \leq 16$$
$$-u_1 \qquad\quad +u_4 \qquad \leq 20$$
$$-u_1 \qquad\qquad\quad +u_5 \leq 22$$
$$-u_2 +u_3 \qquad\qquad \leq 18$$
$$-u_2 \quad\; +u_4 \qquad \leq 16$$
$$-u_2 +u_5 \qquad\qquad \leq 14$$
$$u_1 \geq 0,\, u_2 \geq 0, ..., u_5 \geq 0$$

The initial simplex tableau is

u_1	u_2	u_3	u_4	u_5	x_1	x_2	x_3	x_4	x_5	x_6	P	Const.
-1	0	1	0	0	1	0	0	0	0	0	0	16
-1	0	0	1	0	0	1	0	0	0	0	0	20
-1	0	0	0	1	0	0	1	0	0	0	0	22
0	-1	1	0	0	0	0	0	1	0	0	0	18
0	-1	0	1	0	0	0	0	0	1	0	0	16
0	-1	0	0	1	0	0	0	0	0	1	0	14
800	600	-500	-400	-400	0	0	0	0	0	0	1	0

Using the sequence of row operations
1. $R_4 - R_1$, $R_7 + 500R_1$ 2. $R_2 - R_5$, $R_7 + 400R_5$ 3. $R_3 - R_6$, $R_7 + 400R_6$
4. $R_3 - R_2$, $R_4 + R_2$, $R_5 + R_2$, $R_6 + R_2$, $R_7 + 200R_2$
we obtain the final tableau

u_1	u_2	u_3	u_4	u_5	x_1	x_2	x_3	x_4	x_5	x_6	P	Const.
-1	0	1	0	0	1	0	0	0	0	0	0	16
-1	1	0	0	0	0	1	0	0	-1	0	0	4
0	0	0	0	0	0	-1	1	0	1	-1	0	4
0	0	0	0	0	-1	1	0	1	-1	0	0	6
-1	0	0	1	0	0	1	0	0	0	0	0	20
-1	0	0	0	1	0	1	0	0	-1	1	0	18
100	0	0	0	0	500	200	0	0	200	400	1	20,800

We find $x_1 = 500$, $x_2 = 200$, $x_3 = 0$, $x_4 = 0$, $x_5 = 200$, $x_6 = 400$, and $C = 20,800$. So the schedule is

Location *I*: 500 to warehouse *A*, 200 to warehouse *B*
Location *II*: 200 to warehouse *B*, 400 to warehouse *C*
Shipping costs: $20,800.

23. The given data may be summarized as follows:

	Orange Juice	Grapefruit Juice
Vitamin A	60 I.U.	120 I.U.
Vitamin C	16 I.U.	12 I.U.
Calories	14	11

Suppose x ounces of orange juice and y ounces of pink-grapefruit juice are required for each glass of the blend. Then the problem is

Minimize $C = 14x + 11y$ subject to
$$60x + 120y \geq 1200$$
$$16x + 12y \geq 200$$
$$x \geq 0, y \geq 0$$

To construct the dual problem, we first write down the tableau

x	y	*Const.*
60	120	1200
16	12	200
14	11	

Then obtain the following by interchanging rows and columns:

u	v	*Const.*
60	16	14
120	12	11
1200	200	

From this table we construct the dual problem:

Maximize $P = 1200u + 200v$ subject to

$$60u + 16v \leq 14$$
$$120u + 12v \leq 11$$
$$u \geq 0, v \geq 0$$

The initial tableau is

u	v	x	y	P	Const.
60	16	1	0	0	14
120	12	0	1	0	11
−1200	−200	0	0	1	0

Using the following sequence of row operations,

1. $\frac{1}{120}R_2$　2. $R_1 - 60R_2$, $R_3 + 1200R_2$　3. $\frac{1}{10}R_1$　4. $R_2 - \frac{1}{10}R_1$, $R_3 + 80R_1$

we obtain the final tableau

u	v	x	y	P	Const.
0	1	$\frac{1}{10}$	$-\frac{1}{20}$	0	$\frac{17}{20}$
0	0	$-\frac{1}{100}$	$\frac{1}{75}$	0	$\frac{1}{150}$
0	0	8	6	1	178

We conclude that the owner should use 8 ounces of orange juice and 6 ounces of grapefruit juice per glass of the blend for a minimal calorie count of 178.

25.　True. To maximize P, one maximizes -C. Since the minimization problem has a unique solution, the negative of that solution is the solution of the maximization problem.

USING TECHNOLOGY EXERCISES 4.2, page 262

1.　$x = \frac{4}{3}$, $y = \frac{10}{3}$, $z = 0$, and $C = \frac{14}{3}$

3.　$x = 0.9524$, $y = 4.2857$, $z = 0$, and $C = 6.0952$.

EXERCISES 4.3, page 275

1.　Maximize $C = -P = -2x + 3y$ subject to

$$-3x - 5y \le -20$$
$$3x + y \le 16$$
$$-2x + y \le 1$$
$$x \ge 0, y \ge 0$$

3. Maximize $P = -C = -5x - 10y - z$ subject to
$$-2x - y - z \le -4$$
$$-x - 2y - 2z \le -2$$
$$2x + 4y + 3z \le 12$$
$$x \ge 0, \ y \ge 0, \text{ and } z \ge 0$$

5. We set up the tableau and solve the problem using the simplex method:

x	y	u	v	P	Const	Ratio
2	5	1	0	0	20	4
1	(-5)	0	1	0	-5	1
-1	-2	0	0	1	0	

p.r.→ (row 2) $\xrightarrow{-\frac{1}{5}R_2}$

p.c. (column y)

x	y	u	v	P	Const
2	5	1	0	0	20
$-\frac{1}{5}$	1	0	$-\frac{1}{5}$	0	1
-1	-2	0	0	1	0

$\xrightarrow[R_3+2R_2]{R_1-5R_2}$

x	y	u	v	P	Const	Ratio
(3)	0	1	1	0	15	5
$-\frac{1}{5}$	1	0	$-\frac{1}{5}$	0	1	--
$-\frac{7}{5}$	0	0	$-\frac{2}{5}$	1	2	

$\xrightarrow{\frac{1}{3}R_1}$

x	y	u	v	P	Const
1	0	$\frac{1}{3}$	$\frac{1}{3}$	0	5
$-\frac{1}{5}$	1	0	$-\frac{1}{5}$	0	1
$-\frac{7}{5}$	0	0	$-\frac{2}{5}$	1	2

$\xrightarrow[R_3+\frac{7}{5}R_1]{R_2+\frac{1}{5}R_1}$

x	y	u	v	P	Const
1	0	$\frac{1}{3}$	$\frac{1}{3}$	0	5
0	1	$\frac{1}{15}$	$-\frac{2}{15}$	0	2
0	0	$\frac{7}{15}$	$\frac{1}{15}$	1	9

The maximum value of P is 9 when $x = 5$ and $y = 2$.

7. We first rewrite the problem as a maximization problem with inequality constraints using $\leq$, obtaining the following equivalent problem:

$$\text{Maximize } P = -C = 2x - y \text{ subject to}$$
$$x + 2y \leq 6$$
$$3x + 2y \leq 12$$
$$x \geq 0, y \geq 0$$

Following the procedure outlined for nonstandard problems, we have

	x	y	u	v	P	Const		Ratio	
	1	2	1	0	0	6		6	$\frac{1}{3}R_2$
p.r. →	③	2	0	1	0	12		4	$\longrightarrow$
	−2	1	0	0	1	0			

p.c. ↑ (under x column)

x	y	u	v	P	Const	
1	2	1	0	0	6	$R_1 - R_2$
1	$\frac{2}{3}$	0	$\frac{1}{3}$	0	4	$R_3 + 2R_2$
−2	1	0	0	1	0	$\longrightarrow$

x	y	u	v	P	Const
0	$\frac{4}{3}$	1	$-\frac{1}{3}$	0	2
1	$\frac{2}{3}$	0	$\frac{1}{3}$	0	4
0	$\frac{7}{3}$	0	$\frac{2}{3}$	1	8

We conclude that C attains a minimum value of -8 when $x = 4$ and $y = 0$.

9. Using the simplex method we have

x	y	u	v	P	Const		Ratio	
1	3	1	0	0	6		6	$-\frac{1}{2}R_2$
p.r.→ ⟨-2⟩	3	0	1	0	-6		3	$\longrightarrow$
-1	-4	0	0	1	0			

↑
p.c.

x	y	u	v	P	Const	
1	3	1	0	0	6	$R_1 - R_2$
1	$-\frac{3}{2}$	0	$-\frac{1}{2}$	0	3	$R_3 + R_2$
-1	-4	0	0	1	0	$\longrightarrow$

x	y	u	v	P	Const	
0	⟨$\frac{9}{2}$⟩	1	$\frac{1}{2}$	0	3	$\frac{2}{9}R_1$
1	$-\frac{3}{2}$	0	$-\frac{1}{2}$	0	3	$\longrightarrow$
0	$-\frac{11}{2}$	0	$-\frac{1}{2}$	1	3	

x	y	u	v	P	Const	
0	1	$\frac{2}{9}$	$\frac{1}{9}$	0	$\frac{2}{3}$	$R_2 + \frac{3}{2}R_1$
1	$-\frac{3}{2}$	0	$-\frac{1}{2}$	0	3	$R_3 + \frac{11}{2}R_1$
0	$-\frac{11}{2}$	0	$-\frac{1}{2}$	1	3	$\longrightarrow$

x	y	u	v	P	Const
0	1	$\frac{2}{9}$	$\frac{1}{9}$	0	$\frac{2}{3}$
1	0	$\frac{1}{3}$	$-\frac{1}{3}$	0	4
0	0	$\frac{11}{9}$	$\frac{1}{9}$	1	$\frac{20}{3}$

We conclude that P attains a maximum value of 20/3, when $x = 4$ and $y = 2/3$.

11. We rewrite the problem as

Maximize $P = x + 2y$ subject to
$$2x + 3y \le 12$$
$$-x + 3y \le 3$$
$$-x + 3y \ge 3$$
$$x \ge 0, y \ge 0$$

The initial tableau is

x	y	u	v	w	P	Const.
2	3	1	0	0	0	12
−1	3	0	1	0	0	3
1	−3	0	0	1	0	−3
−1	−2	0	0	0	1	0

Using the following sequence of row operations

1. $-\frac{1}{3}R_3$ 2. $R_1 - 3R_3$, $R_2 - 3R_3$, $R_4 + 2R_3$ 3. $\frac{1}{3}R_1$
4. $R_3 + \frac{1}{3}R_1$, $R_4 + \frac{5}{3}R_1$ 5. $R_1 - \frac{1}{3}R_2$, $R_3 + \frac{2}{9}R_2$, $R_4 + \frac{1}{9}R_2$

we obtain the final tableau

x	y	u	v	w	P	Const.
1	0	$\frac{1}{3}$	$-\frac{1}{3}$	0	0	3
0	0	0	1	1	0	0
0	1	$\frac{1}{9}$	$\frac{2}{9}$	0	0	2
0	0	$\frac{5}{9}$	$\frac{1}{9}$	0	1	7

We conclude that P attains a maximum value of 7 when $x = 3$ and $y = 2$.

13. We rewrite the problem as

$$\text{Maximize } P = 5x + 4y + 2z \text{ subject to}$$
$$x + 2y + 3z \le 24$$
$$-x + y - z \le -6$$
$$x \ge 0,\ y \ge 0,\ z \ge 0$$

The initial tableau is

x	y	z	u	v	P	Const.
1	2	3	1	0	0	24
−1	1	−1	0	1	0	−6
−5	−4	−2	0	0	1	0

Using the following sequence of row operations
1. $-R_2$ 2. $R_1 - 2R_2$, $R_3 + 5R_2$ 3. $\frac{1}{3}R_1$ 4. $R_2 + R_1$, $R_3 + 9R_1$
5. $3R_1$ 6. $R_2 + \frac{2}{3}R_1$; $R_3 + 2R_1$

we obtain the final tableau

x	y	z	u	v	P	Const.
0	3	2	1	1	0	18
1	2	3	1	0	0	24
0	6	13	5	0	1	120

from which we deduce that P attains a maximum value of 120 when $x = 24$, $y = 0$, and $z = 0$.

15. The problem is to maximize $P = -C = -x + 2y - z$ subject to the given constraints. The initial tableau is

x	y	z	u	v	w	P	Const.
1	-2	3	1	0	0	0	10
2	1	-2	0	1	0	0	15
2	1	3	0	0	1	0	20
1	-2	1	0	0	0	1	0

Using the following sequence of row operations,

1. $R_1 + 2R_2$, $R_3 - R_2$, $R_4 + 2R_2$ 2. $\frac{1}{5}R_3$ 3. $R_1 + R_3$, $R_2 + 2R_3$, $R_4 + 3R_3$

we obtain the final tableau

x	y	z	u	v	w	P	Const.
5	0	0	1	$\frac{9}{5}$	$\frac{1}{5}$	0	41
2	1	0	0	$\frac{3}{5}$	$\frac{2}{5}$	0	17
0	0	1	0	$-\frac{1}{5}$	$\frac{1}{5}$	0	1
5	0	0	0	$\frac{7}{5}$	$\frac{3}{5}$	1	33

We conclude that C attains a minimum value of -33 when $x = 0$, $y = 17$, $z = 1$, and $C = -P = -33$.

17. Rewriting the third constraint as $-x + 2y - z \le -4$, we obtain the following initial tableau

x	y	z	u	v	w	P	Const.
1	2	3	1	0	0	0	28
2	3	-1	0	1	0	0	6
-1	2	-1	0	0	1	0	-4
-2	-1	-1	0	0	0	1	0

Using the following sequence of row operations,

1. $\frac{1}{2}R_2$ 2. $R_1 - R_2$, $R_3 + R_2$, $R_4 + 2R_2$ 3. $-\frac{2}{3}R_3$
4. $R_1 - \frac{7}{2}R_3$, $R_2 + \frac{1}{2}R_3$, $R_4 + 2R_3$ 5. $\frac{3}{26}R_1$ 6. $R_2 - \frac{1}{3}R_1$; $R_3 + \frac{7}{3}R_1$, $R_4 + \frac{8}{3}R_1$
7. $\frac{26}{7}R_1$ 8. $R_2 + \frac{11}{26}R_1$, $R_3 + \frac{1}{26}R_1$, $R_4 + \frac{8}{13}R_1$

we obtain the final tableau

x	y	z	u	v	w	P	Const.
0	$\frac{26}{7}$	0	$\frac{3}{7}$	$\frac{2}{7}$	1	0	$\frac{68}{7}$
1	$\frac{11}{7}$	0	$\frac{1}{7}$	$\frac{3}{7}$	1	0	$\frac{46}{7}$
0	$\frac{1}{7}$	1	$\frac{2}{7}$	$-\frac{1}{7}$	0	0	$\frac{50}{7}$
0	$\frac{16}{7}$	0	$\frac{4}{7}$	$\frac{5}{7}$	0	1	$\frac{142}{7}$

from which we deduce that P attains a maximum value of 142/7 when $x = 46/7$, $y = 0$, and $z = 50/7$.

19. Rewriting the third constraint $(2x + y + z = 10)$ in the form
$$2x + y + z \geq 10 \quad \text{and} \quad -2x - y - z \leq -10,$$
we obtain the following initial tableau.

x	y	z	t	u	v	w	P	Const.
1	2	1	1	0	0	0	0	20
3	1	0	0	1	0	0	0	30
2	1	1	0	0	1	0	0	10
-2	-1	-1	0	0	0	1	0	-10
-1	-2	-3	0	0	0	0	1	0

Using the following sequence of row operations
1. $-R_4$ 2. $R_1 - R_4$, $R_3 - R_4$, $R_5 + 3R_4$ 3. $R_1 - R_3$, $R_4 + R_3$, $R_5 + 3R_3$
we obtain the final tableau

x	y	z	t	u	v	w	P	Const.
-1	1	0	1	0	-1	0	0	10
3	1	0	0	1	0	0	0	30
0	0	0	0	0	1	1	0	0
2	1	1	0	0	1	0	0	10
5	1	0	0	0	3	0	1	30

We conclude that P attains a maximum value of 30 when $x = 0$, $y = 0$, and $z = 10$.

21. Let x and y denote the number of acres of crops A and B, respectively to be planted.
Then the problem is

Maximize $P = 150x + 200y$ subject to the constraints

$$x + y \leq 150$$
$$40x + 60y \leq 7400$$
$$20x + 25y \leq 3300$$
$$x \geq 80$$
$$x \geq 0, y \geq 0$$

Using the simplex method, we obtain

x	y	u	v	w	z	P	Const.	Ratio
1	1	1	0	0	0	0	150	150
40	60	0	1	0	0	0	7400	185
20	25	0	0	1	0	0	3300	165
(-1)	0	0	0	0	1	0	-80	--
-150	-200	0	0	0	0	1	0	

$\xrightarrow{-R_4}$

x	y	u	v	w	z	P	Const.
1	1	1	0	0	0	0	150
40	60	0	1	0	0	0	7400
20	25	0	0	1	0	0	3300
1	0	0	0	0	-1	0	80
-150	-200	0	0	0	0	1	0

$\begin{array}{c} R_1 - R_4 \\ R_2 - 40R_4 \\ R_3 - 20R_4 \\ R_5 + 150R_4 \end{array}$ $\longrightarrow$

x	y	u	v	w	z	P	Const.	Ratio
0	1	1	0	0	1	0	70	70
0	60	0	1	0	40	0	4200	70
0	(25)	0	0	1	20	0	1700	68
1	0	0	0	0	-1	0	80	--
0	-200	0	0	0	-150	1	12,000	

$\xrightarrow{\frac{1}{25}R_3}$

x	y	u	v	w	z	P	Const.
0	1	1	0	0	1	0	70
0	60	0	1	0	40	0	4200
0	1	0	0	$\frac{1}{25}$	$\frac{4}{5}$	0	68
1	0	0	0	0	-1	0	80
0	-200	0	0	0	-150	1	12,000

$$R_1 - R_3$$
$$R_2 - 60R_3$$
$$R_5 + 200R_3 \longrightarrow$$

x	y	u	v	w	z	P	Const.
0	0	1	0	$-\frac{1}{25}$	$\frac{1}{5}$	0	2
0	0	0	1	$-\frac{12}{5}$	-8	0	120
0	1	0	0	$\frac{1}{25}$	$\frac{4}{5}$	0	68
1	0	0	0	0	-1	0	80
0	0	0	0	8	10	1	25,600

We conclude that the farmer should plant 80 acres of crop A and 68 acres of crop B to realize a maximum profit of $25,600.

23. Let x and y denote the amount (in dollars) invested in company A and company B, respectively. Then the problem is

$$\text{Maximize } P = 0.08x + 0.06y \text{ subject to}$$
$$-x + 3y \le 0$$
$$y \ge 10,000,000$$
$$x + y = 60,000,000$$
$$x \ge 0,\ y \ge 0$$

Substituting $x = 60,000,000 - y$ into the first equation and the first and second inequalities, we have

$$\text{Maximize } P = 0.08(60,000,000 - y) + 0.06y$$
$$= 4,800,000 - 0.02y \text{ subject to}$$
$$y \le 15,000,000$$
$$y \ge 10,000,000$$
$$x \ge 0,\ y \ge 0.$$

Using the simplex method, we have

y	u	v	P	Const.
1	1	0	0	15,000,000
(−1)	0	1	0	−10,000,000
0.02	0	0	1	4,800,000

p.r. → points to the (-1) row, p.c. ↑ points to the y column. $\xrightarrow{-R_2}$

y	u	v	P	Const.
1	1	0	0	15,000,000
1	0	−1	0	10,000,000
0.02	0	0	1	4,800,000

$\xrightarrow[R_5 - 0.02R_2]{R_1 - R_2}$

y	u	v	P	Const.
0	1	1	0	5,000,000
1	0	−1	0	10,000,000
0	0	0.02	1	4,600,000

We conclude that the bank should extend $50 million in home loans, $10 million of commercial-development loans to attain a maximum return of $4.6 million.

25. Let x, y, and z denote the number of units of products A, B, and C manufactured by the company. Then the linear programming problem is

Maximize $P = 18z + 12y + 15z$ subject to
$$2x + y + 2z \le 900$$
$$3x + y + 2z \le 1080$$
$$2x + 2y + z \le 840$$
$$x - y + z \le 0$$
$$x \ge 0,\ y \ge 0,\ z \ge 0$$

The initial tableau is

x	y	z	t	u	v	w	P	Const.
2	1	2	1	0	0	0	0	900
3	1	2	0	1	0	0	0	1080
2	2	1	0	0	1	0	0	840
1	−1	1	0	0	0	1	0	0
−18	−12	−15	0	0	0	0	1	0

Using the following sequence of row operations,

1. $R_1 - 2R_4$, $R_2 - 3R_4$, $R_3 - 2R_4$, $R_5 + 18R_4$ 2. $\frac{1}{4}R_3$

3. $R_1 - 3R_3$, $R_2 - 4R_3$, $R_4 + R_3$, $R_5 + 30R_3$ 4. $\frac{4}{3}R_4$

5. $R_1 - \frac{3}{4}R_4$, $R_3 + \frac{1}{4}R_4$, $R_5 + \frac{9}{2}R_4$

we obtain the final tableau

x	y	z	t	u	v	w	P	Const.
−1	0	0	1	0	−1	−1	0	60
0	0	0	0	1	−1	−1	0	240
$\frac{1}{3}$	1	0	0	0	$\frac{1}{3}$	−$\frac{1}{3}$	0	280
$\frac{4}{3}$	0	1	0	0	$\frac{1}{3}$	$\frac{2}{3}$	0	280
−6	0	0	0	0	9	6	1	7560

and conclude that the company should produce 0 units of product A, 280 units of product B, and 280 units of product C to realize a maximum profit of $7,560.

27. Let x denote the number of ounces of food A and y denote the number of ounces of food B used in the meal. Then the problem is to minimize the amount of cholesterol in the meal. Thus, the linear programming problem is

Maximize $P = -C = -2x - 5y$ subject to

$$30x + 25y \geq 400$$

$$x + \tfrac{1}{2}y \geq 10$$

$$2x + 5y \geq 40$$

$$x \geq 0, y \geq 0$$

The initial tableau is

x	y	u	v	w	C	Const.
−30	−25	1	0	0	0	−400
−1	−$\frac{1}{2}$	0	1	0	0	−10
−2	−5	0	0	1	0	−40
2	5	0	0	0	1	0

Using the following sequence of row operations

1. $-R_2$ 2. $R_1 + 30R_2$; $R_3 + 2R_2$; $R_4 - 2R_2$ 3. $-\frac{1}{4}R_3$

4. $R_1 + 10R_2$; $R_2 - \frac{1}{2}R_3$; $R_4 - 4R_3$ 5. $-\frac{1}{25}R_1$ 6. $R_2 + \frac{5}{4}R_1$; $R_3 - \frac{1}{2}R_1$

we obtain the final tableau

x	y	u	v	w	C	Const.
0	0	$-\frac{1}{25}$	1	$\frac{1}{10}$	0	2
1	0	$-\frac{1}{20}$	1	$\frac{1}{4}$	0	10
0	1	$\frac{1}{50}$	0	$-\frac{3}{10}$	0	4
0	0	0	0	1	1	-40

Thus, the minimum content of cholesterol is 40mg when 10 ounces of food A and 4 ounces of food B are used.(Since the u-column is not in unit form, we see that the problem has multiple solutions.)

CHAPTER 4 REVIEW EXERCISES, page 280

1. This is a regular linear programming problem. Using the simplex method with u and v as slack variables, we obtain the following sequence of tableaus:

	x	y	u	v	P	Const	Ratio	
p.r.→	1	③	1	0	0	15	5	$\frac{1}{3}R_1$ →
	4	1	0	1	0	16	16	
	-3	-4	0	0	1	0		

↑
p.c.

x	y	u	v	P	Const.	
$\frac{1}{3}$	1	$\frac{1}{3}$	0	0	5	R_2-R_1
4	1	0	1	0	16	R_3+4R_1 →
-3	-4	0	0	1	0	

	x	y	u	v	P	Const.	
	$\frac{1}{3}$	1	$\frac{1}{3}$	0	0	5	$\frac{3}{11}R_2$ →
p.c.→	$\frac{11}{3}$	0	$-\frac{1}{3}$	1	0	11	
	$-\frac{5}{3}$	0	$\frac{4}{3}$	0	1	20	

↑
p.c.

x	y	u	v	P	Const.	
$\frac{1}{3}$	1	$\frac{1}{3}$	0	0	5	$R_1 - \frac{1}{3}R_2$
1	0	$-\frac{1}{11}$	$\frac{3}{11}$	0	3	$R_3 + \frac{5}{3}R_2$
$-\frac{5}{3}$	0	$\frac{4}{3}$	0	1	20	$\longrightarrow$

x	y	u	v	P	Const.
0	1	$\frac{4}{11}$	$-\frac{1}{11}$	0	4
1	0	$-\frac{1}{11}$	$\frac{3}{11}$	0	3
0	0	$\frac{13}{11}$	$\frac{5}{11}$	1	25

and conclude that $x = 3$, $y = 4$, and $P = 25$.

3. Using the simplex method to solve this regular linear programming problem we have

	x	y	z	u	v	P	Const.	Ratio	
$p.r. \rightarrow$	1	2	③	1	0	0	12	4	$\frac{1}{3}R_1$
	1	−3	2	0	1	0	10	5	$\longrightarrow$
	−2	−3	−5	0	0	1	0		

$$\uparrow \quad p.c.$$

x	y	z	u	v	P	Const.	
$\frac{1}{3}$	$\frac{2}{3}$	1	$\frac{1}{3}$	0	0	4	$R_2 - 2R_1$
1	−3	2	0	1	0	10	$R_3 + 5R_1$
−2	−3	−5	0	0	1	0	$\longrightarrow$

	x	y	z	u	v	P	Const.	Ratio	
	$\frac{1}{3}$	$\frac{2}{3}$	1	$\frac{1}{3}$	0	0	4	12	$3R_2$
$p.r. \rightarrow$	⑴⁄₃	$-\frac{13}{3}$	0	$-\frac{2}{3}$	1	0	2	6	$\longrightarrow$
	$-\frac{1}{3}$	$\frac{1}{3}$	0	$\frac{5}{3}$	0	1	20		

$$\uparrow \quad p.c.$$

x	y	z	u	v	P	Const.	
$\frac{1}{3}$	$\frac{2}{3}$	1	$\frac{1}{3}$	0	0	4	$R_1 - \frac{1}{3}R_2$
1	−13	0	−2	3	0	6	$R_3 + \frac{1}{3}R_2$
$-\frac{1}{3}$	$\frac{1}{3}$	0	$\frac{5}{3}$	0	1	20	$\longrightarrow$

	x	y	z	u	v	P	Const.	Ratio	
$p.r.\rightarrow$	0	⑤	1	1	-1	0	2	$2/5$	$\frac{1}{5}R_1$
	1	-13	0	-2	3	0	6	--	
	0	-4	0	1	1	1	22		

$$\uparrow$$
$$p.c.$$

x	y	z	u	v	P	Const.
0	1	$\frac{1}{5}$	$\frac{1}{5}$	$-\frac{1}{5}$	0	$\frac{2}{5}$
0	-13	0	-2	3	0	6
0	-4	0	1	1	1	22

$$\xrightarrow{\begin{array}{c}R_2+13R_1\\R_3+4R_1\end{array}}$$

x	y	z	u	v	P	Const.
0	1	$\frac{1}{5}$	$\frac{1}{5}$	$-\frac{1}{5}$	0	$\frac{2}{5}$
1	0	$\frac{13}{5}$	$\frac{3}{5}$	$\frac{2}{5}$	0	$\frac{56}{5}$
0	0	$\frac{4}{5}$	$\frac{9}{5}$	$\frac{1}{5}$	1	$\frac{118}{5}$

We conclude that the P attains a maximum value of 23.6 when $x = 11.2$, $y = 0.4$ and $z = 0$.

5. We first write the tableau

x	y	Const.
2	3	6
2	1	4
3	2	

Then obtain the following by interchanging rows and columns:

u	v	Const.
2	2	3
3	1	2
6	4	

From this table we construct the dual problem:
Maximize the objective function $P = 6u + 4v$ subject to the constraints
$$2u + 2v \leq 3$$
$$3u + v \leq 2$$
$$u \geq 0, v \geq 0$$

Using the simplex method, we have

$$
\begin{array}{ccccc|c}
u & v & x & y & P & \textit{Const} \\
\hline
2 & 2 & 1 & 0 & 0 & 3 \\
\boxed{3} & 1 & 0 & 1 & 0 & 2 \\
-6 & -4 & 0 & 0 & 1 & 0
\end{array}
$$

p.r. $\rightarrow$ points to the 3 in row 2; p.c. points to the u column.

	Ratio	
	$3/2$	$\tfrac{1}{3}R_2$
	$2/3$	$\longrightarrow$

$$
\begin{array}{ccccc|c}
u & v & x & y & P & \textit{Const} \\
\hline
2 & 2 & 1 & 0 & 0 & 3 \\
1 & \tfrac{1}{3} & 0 & \tfrac{1}{3} & 0 & \tfrac{2}{3} \\
-6 & -4 & 0 & 0 & 1 & 0
\end{array}
\quad
\begin{array}{l} R_1 - 2R_2 \\ R_3 + 6R_2 \\ \longrightarrow \end{array}
\quad
\begin{array}{ccccc|c}
u & v & x & y & P & \textit{Const} \\
\hline
0 & \boxed{\tfrac{4}{3}} & 1 & -\tfrac{2}{3} & 0 & \tfrac{5}{3} \\
1 & \tfrac{1}{3} & 0 & \tfrac{1}{3} & 0 & \tfrac{2}{3} \\
0 & -2 & 0 & 2 & 1 & 4
\end{array}
$$

Ratio	
$5/4$	$\tfrac{3}{4}R_1$
2	$\longrightarrow$

$$
\begin{array}{ccccc|c}
u & v & x & y & P & \textit{Const} \\
\hline
0 & 1 & \tfrac{3}{4} & -\tfrac{1}{2} & 0 & \tfrac{5}{4} \\
1 & \tfrac{1}{3} & 0 & \tfrac{1}{3} & 0 & \tfrac{2}{3} \\
0 & -2 & 0 & 2 & 1 & 4
\end{array}
\quad
\begin{array}{l} R_2 - \tfrac{1}{3}R_1 \\ R_3 + 2R_1 \\ \longrightarrow \end{array}
\quad
\begin{array}{ccccc|c}
u & v & x & y & P & \textit{Const} \\
\hline
0 & 1 & \tfrac{3}{4} & -\tfrac{1}{2} & 0 & \tfrac{5}{4} \\
1 & 0 & -\tfrac{1}{4} & \tfrac{1}{2} & 0 & \tfrac{1}{4} \\
0 & 0 & \tfrac{3}{2} & 1 & 1 & \tfrac{13}{2}
\end{array}
$$

Therefore, C attains a minimum value of $13/2$ when $x = 3/2$ and $y = 1$.

7. We first write the tableau

$$
\begin{array}{ccc|c}
x & y & z & \textit{Const.} \\
\hline
3 & 2 & 1 & 4 \\
1 & 1 & 3 & 6 \\
24 & 18 & 24 &
\end{array}
$$

Then obtain the following by interchanging rows and columns:

$$
\begin{array}{cc|c}
u & v & \textit{Const.} \\
\hline
3 & 1 & 24 \\
2 & 1 & 18 \\
1 & 3 & 24 \\
\hline
4 & 6 &
\end{array}
$$

From this table we construct the dual problem:
Maximize the objective function $P = 4u + 6v$ subject to

$$3u + v \le 24$$
$$2u + v \le 18$$
$$u + 3v \le 24$$
$$u \ge 0, v \ge 0$$

The initial tableau is

u	v	x	y	z	P	Const.
3	1	1	0	0	0	24
2	1	0	1	0	0	18
1	3	0	0	1	0	24
−4	−6	0	0	0	1	0

Using the following sequence of row operations

1. $\frac{1}{3}R_3$ 2. $R_1 - R_3$, $R_2 - R_3$, $R_4 + 6R_3$ 3. $\frac{3}{8}R_1$ 4. $R_2 - \frac{2}{3}R_1$, $R_3 - \frac{1}{3}R_1$, $R_4 + 2R_1$

we obtain the final tableau

u	v	x	y	z	P	Const.
1	0	$\frac{3}{8}$	0	$-\frac{1}{8}$	0	6
0	0	$-\frac{5}{8}$	1	$-\frac{1}{8}$	0	0
0	1	$-\frac{1}{8}$	0	$\frac{3}{8}$	0	6
0	0	$\frac{3}{4}$	0	$\frac{7}{4}$	0	60

We conclude that C attains a minimum value of 60 when $x = 3/4$, $y = 0$, and $z = 7/4$.

9. Rewriting the problem, we have

Maximize $P = 3x - 4y$ subject to
$$x + y \le 45$$
$$-x + 2y \le -10$$
$$x \ge 0, y \ge 0$$

Using the simplex method, we have

x	y	u	v	P	Const
1	1	1	0	0	45
(−1)	2	0	1	0	−10
−3	4	0	0	1	0

Ratio
45
10

$\xrightarrow{-R_2}$

x	y	u	v	P	Const
1	1	1	0	0	45
1	−2	0	−1	0	10
−3	4	0	0	1	0

$\xrightarrow[R_3 + 3R_2]{R_1 - R_2}$

x	y	u	v	P	Const
0	3	1	1	0	35
1	-2	0	-1	0	10
0	-2	0	-3	1	30

$$\xrightarrow[R_3+3R_1]{R_2+R_1}$$

x	y	u	v	P	Const
0	3	1	1	0	35
1	1	1	0	0	45
0	7	3	0	1	135

We conclude that P attains a maximum value of 135 when $x = 45$ and $y = 0$.

11. We first write the problem in the form

Maximize $P = 2x + 3y$ subject to
$$2x + 5y \le 20$$
$$x - 5y \le -5$$
$$x \ge 0,\ y \ge 0$$

The initial tableau is

x	y	u	v	P	Const
2	5	1	0	0	20
1	-5	0	1	0	-5
-2	-3	0	0	1	0

Using the sequence of row operations

1. $\frac{1}{5}R_1$ 2. $R_2 + 5R_1,\ R_3 + 3R_1$ 3. $\frac{1}{3}R_2$ 4. $R_1 - \frac{2}{5}R_2,\ R_3 + \frac{4}{5}R_2$

we obtain the final tableau

x	y	u	v	P	Const
0	1	$\frac{1}{15}$	$-\frac{2}{15}$	0	2
1	0	$\frac{1}{3}$	$\frac{1}{3}$	0	5
0	0	$\frac{13}{15}$	$\frac{4}{15}$	1	16

We conclude that P attains a maximum value of 16 when $x = 5$ and $y = 2$.

13. Let x, y, and z denote the number of units of products A, B, and C made, respectively. Then the problem is to maximize the profit
$$P = 4x + 6y + 8z \text{ subject to}$$
$$9x + 12y + 18z \le 360$$
$$6x + 6y + 10z \le 240$$
$$x \ge 0,\ y \ge 0,\ z \ge 0$$

The initial tableau is

x	y	z	u	v	P	Const.
9	12	18	1	0	0	360
6	6	10	0	1	0	240
-4	-6	-8	0	0	1	0

Using the sequence of row operations

1. $\frac{1}{18}R_1$ 2. $R_2 - 10R_1$ 3. $R_3 + 8R_1$ 4. $\frac{3}{2}R_1$ 5. $R_2 + \frac{2}{3}R_1$, $R_3 + \frac{2}{3}R_1$

we obtain the final tableau

x	y	z	u	v	P	Const.
$\frac{3}{4}$	1	$\frac{3}{2}$	$\frac{1}{12}$	0	0	30
$\frac{3}{2}$	0	1	$-\frac{1}{2}$	1	0	60
$\frac{1}{2}$	0	1	$\frac{1}{2}$	0	1	180

and conclude that the company should produce 0 units of product A, 30 units of product B, and 0 units of product C to realize a maximum profit of $180.

15. Let x, y, and z, denote the amount invested in stocks, bonds, and money-market funds, respectively. Then the problem is

$$\text{Maximize } P = 0.15x + 0.10y + 0.08z \text{ subject to}$$
$$x + y + z \le 200,000$$
$$-z \le -50,000$$
$$x - y + z \le 0$$
$$x \ge 0,\ y \ge 0,\ z \ge 0$$

The initial tableau is

x	y	z	u	v	w	P	Const.
1	1	1	1	0	0	0	200,000
0	0	-1	0	1	0	0	-50,000
1	-1	1	0	0	1	0	0
-0.15	-0.10	-0.08	0	0	0	1	0

Using the following sequence of row operations,

1. $-R_2$ 2. $R_1 - R_2$, $R_3 - R_2$, $R_4 + 0.08R_2$ 3. $-R_3$ 4. $R_1 - R_3$, $R_4 + \frac{1}{10}R_3$
5. $\frac{1}{2}R_1$ 6. $R_3 + R_1$, $R_4 + \frac{1}{4}R_1$

we obtain the final tableau

x	y	z	u	v	w	P	Const.
1	0	0	$\frac{1}{2}$	1	$\frac{1}{2}$	0	50,000
0	0	1	0	−1	0	0	50,000
0	1	0	$\frac{1}{2}$	0	$-\frac{1}{2}$	0	100,000
0	0	0	$\frac{1}{8}$	$\frac{7}{100}$	$\frac{1}{40}$	1	21,500

and we conclude that Sandra should invest $50,000 in stocks, $100,000 in bonds, and $50,000 in money-market funds to realize a maximum return of $21,500 per year on her investments.

CHAPTER 5

EXERCISES 5.1, page 293

1. The interest is given by $I = (500)(2)(0.08) = 80$, or $80.
    The accumulated amount is $500 + 80$, or $580.

3. The interest is given by $I = (800)(0.06)(0.75) = 36$, or $36.
    The accumulated amount is $800 + 36$, or $836.

5. We are given that $A = 1160$, $t = 2$, and $r = 0.08$, and we are asked to find P. Since
    $$A = P(1 + rt)$$
    we see that
    $$P = \frac{A}{1+rt} = \frac{1160}{1+(0.08)(2)} = 1000, \text{ or } \$1000.$$

7. We use the formula $I = Prt$ and solve for t when $I = 20$, $P = 1000$, and $r = 0.05$.
    Thus,
    $$20 = 1000(0.05)(\frac{t}{365})$$
    and
    $$t = \frac{365(20)}{50} = 146, \quad \text{or 146 days.}$$

9. We use the formula $A = P(1 + rt)$ with $A = 1075$, $P = 1000$, $t = 0.75$, and solve for r.
    Thus,
    $$1075 = 1000(1 + 0.75r)$$
    $$75 = 750r$$
    or
    $$r = 0.10.$$
    Therefore, the interest rate is 10 percent per year.

11. $A = 1000(1 + 0.07)^8 \approx 1718.19$, or $1718.19.

13. $A = 2500\left(1+\dfrac{0.07}{2}\right)^{20} \approx 4974.47$, or $4974.47.

15. $A = 12000\left(1+\dfrac{0.08}{4}\right)^{42} \approx 27,566.93$, or $27,566.93.

17. $A = 150,000\left(1+\dfrac{0.14}{12}\right)^{48} \approx 261,751.04$, or \$261,751.04.

19. $A = 150,000\left(1+\dfrac{0.12}{365}\right)^{1095} = 214,986.69$, or \$214,986.69.

21. Using the formula
$$r_{\textit{eff}} = \left(1+\frac{r}{m}\right)^{m} - 1$$
with $r = 0.10$ and $m = 2$, we have
$$r_{\textit{eff}} = \left(1+\frac{0.10}{2}\right)^{2} - 1 = 0.1025, \quad \text{or } 10.25 \text{ percent}..$$

23. Using the formula
$$r_{\textit{eff}} = \left(1+\frac{r}{m}\right)^{m} - 1$$
with $r = 0.08$ and $m = 12$, we have
$$r_{\textit{eff}} = \left(1+\frac{0.08}{12}\right)^{12} - 1 \approx 0.08300, \quad \text{or } 8.3 \text{ percent per year.}$$

25. The present value is given by
$$P = 40,000\left(1+\frac{0.08}{2}\right)^{-8} \approx 29,227.61, \quad \text{or } \$29,227.61.$$

27. The present value is given by
$$P = 40,000\left(1+\frac{0.07}{12}\right)^{-48} \approx 30,255.95, \quad \text{or } \$30,255.95.$$

29. Think of \$300 as the principal and \$306 as the accumulated amount at the end of 30 days. If r denotes the simple interest rate per annum, then we have $P = 300$, $A = 306$, $n = 1/12$, and we are required to find r. Using (8b) we have
$$306 = 300\left(1+\frac{r}{12}\right) = 300 + r\left(\frac{300}{12}\right)$$
and $\quad r = \left(\dfrac{12}{300}\right)6 = 0.24$, or 24 percent per year.

31. The rate that you would expect to pay is
$$A = 380(1 + 0.08)^5 \approx 558.34, \text{ or } \$558.34 \text{ per day.}$$

33. The amount that they can expect to pay is given by
$$A = 150{,}000(1 + 0.05)^4 \approx 182{,}325.94, \text{ or approximately } \$182{,}326.$$

35. The investment will be worth
$$A = 1.5\left(1 + \frac{0.095}{2}\right)^{20} = 3.794651$$

or approximately $3.8 million dollars.

37. Using the formula
$$P = A\left(1 + \frac{r}{m}\right)^{-mt}$$

we have
$$P = 40{,}000\left(1 + \frac{0.085}{4}\right)^{-20} \approx 26{,}267.49, \text{ or } \$26{,}267.49.$$

39. a. They should set aside
$$P = 100{,}000(1 + 0.085)^{-13} \approx 34{,}626.88, \text{ or } \$34{,}626.88.$$

b. They should set aside
$$P = 100{,}000\left(1 + \frac{0.085}{2}\right)^{-26} \approx 33{,}886.16, \text{ or } \$33{,}886.16.$$

c. They should set aside
$$P = 100{,}000\left(1 + \frac{0.085}{4}\right)^{-52} \approx 33{,}506.76, \text{ or } \$33{,}506.76.$$

41. The present value of the $8000 loan due in 3 years is given by
$$P = 8000\left(1 + \frac{0.10}{2}\right)^{-6} = 5969.72, \text{ or } \$5969.72.$$

The present value of the $15,000 loan due in 6 years is given by
$$P = 15{,}000\left(1 + \frac{0.10}{2}\right)^{-12} = 8352.56, \text{ or } \$8352.56.$$

Therefore, the amount the proprietors of the inn will be required to pay at the end of 5 years is given by
$$A = 14{,}322.28\left(1 + \frac{0.10}{2}\right)^{10} = 23{,}329.48, \text{ or } \$23{,}329.48.$$

43. Using the compound interest formula with $A = 128{,}000$, $P = 100{,}000$ and $t = 6$, we have

$$128{,}000 = 100{,}000(1+R)^6$$
$$(1 + R)^{1/6} = (1.28)^{1/6}$$
$$1 + R = 1.042,$$
$$R = 0.042, \quad \text{or } 4.2 \text{ percent.}$$

45. Let the effective rate of interest be R. Then R satisfies
$$A = P(1 + R)^t$$
or
$$10{,}000 = 6595.37\left(1+\frac{R}{2}\right)^{10}$$

$$1+\frac{R}{2} = (1.51621516)^{1/10}$$
$$= 1.0425$$
and $\qquad R = 0.085$, or 8.5 percent.

47. a. We obtain a family of straight lines with varying slope and P-intercept as P increases. For a fixed rate of interest, the accumulated amount A grows at the rate of Pr units per year starting initially with an amount of $\$P$.
b. We obtain a family of straight lines emanating from the point $(0, P)$ and with varying slope as r increases. For a fixed principal, the accumulated amount A grows at the rate Pr units per year starting initially with an amount of $\$P$.

49. False. Under compound interest $A = P(1+r)^t$ $(m = 1)$, whereas under simple interest, $A = P(1 + rt)$.

51. False. If Susan had gotten annual increases of 5 percent over 5 years, her salary would have been $A = 40{,}000(1+0.05)^5 = 51{,}051.26$, or approximately $\$51{,}051$ after 5 years and not $\$50{,}000$.

USING TECHNOLOGY EXERCISES 5.1, page 297

1. $5872.78   3. $475.49   5. 8.95%/yr   7. 10.20%/yr

9. :PROGRAM: PREVAL
       :Disp "A"
       :Input A
       :Disp "r"

```
:Input r
:Disp "t"
:Input t
:Disp "m"
:Input m
:A(1 + r/m)^(−m*t) → P
:Disp "PRESENT VALUE IS"
:Disp P
```

11. $94,038.74   13. $62,244.96

EXERCISES 5.2, page 306

1. $S = 1000 \left[\dfrac{(1+0.1)^{10} - 1}{0.1} \right] = 15{,}937.42,$ or $15,937.42.

3. $S = 1800 \left[\dfrac{\left(1 + \dfrac{0.08}{4}\right)^{24} - 1}{\dfrac{0.08}{4}} \right] \approx 54{,}759.35,$ or $54,759.35.

5. $S = 600 \left[\dfrac{\left(1 + \dfrac{0.12}{4}\right)^{36} - 1}{\dfrac{0.12}{4}} \right] \approx 37{,}965.57,$ or $37,965.57.

7. $P = 5000 \left[\dfrac{1 - (1+0.08)^{-8}}{0.08} \right] \approx 28{,}733.19,$ or $28,733.19.

9. $P = 4000 \left[\dfrac{1 - (1+0.09)^{-5}}{0.09} \right] \approx 15{,}558.61,$ or $15,558.61.

11. $P = 800 \left[\dfrac{1 - \left(1 + \dfrac{0.12}{4} \right)^{-28}}{\dfrac{0.12}{4}} \right] \approx 15{,}011.29$, or \$15,011.29.

13. She will have

$$S = 1500 \left[\dfrac{(1 + 0.08)^{25} - 1}{0.08} \right] \approx 109{,}658.91, \text{ or } \$109{,}658.91.$$

15. On October 31, Mrs Lynde's account will be worth

$$S = 40 \left[\dfrac{\left(1 + \dfrac{0.07}{12} \right)^{11} - 1}{\dfrac{0.07}{12}} \right] \approx 453.06, \text{ or } \$453.06.$$

One month later, this account will be worth $A = (453.06) \left(1 + \dfrac{0.07}{12} \right) = 455.70$, or \$455.70.

17. The amount in Collin's employee retirement account is given by

$$S = 100 \left[\dfrac{\left(1 + \dfrac{0.07}{12} \right)^{144} - 1}{\dfrac{0.07}{12}} \right] \approx 22{,}469.50, \text{ or } \$22{,}469.50.$$

The amount in Collin's IRA is given by

$$S = 2000 \left[\dfrac{(1 + 0.09)^{8} - 1}{0.09} \right] \approx 22{,}056.95, \text{ or } \$22{,}056.95.$$

Therefore, the total amount in his retirement fund is given by

$$22{,}469.50 + 22{,}056.95 = 44{,}526.45, \text{ or } \$44{,}526.45.$$

19. The equivalent cash payment is given by

$$P = 450\left[\frac{1-\left(1+\frac{0.09}{12}\right)^{-24}}{\frac{0.09}{12}}\right] \approx 9850.12, \text{ or } \$9850.12.$$

21. We use the formula for the present value of an annuity obtaining

$$P = 22\left[\frac{1-\left(1+\frac{0.18}{12}\right)^{-36}}{\frac{0.18}{12}}\right] \approx 608.54, \text{ or } \$608.54.$$

23. With an $800 monthly payment, the present value of their loan would be

$$P = 800\left[\frac{1-\left(1+\frac{0.095}{12}\right)^{-360}}{\frac{0.095}{12}}\right] \approx 95,141.34, \text{ or } \$95,141.34.$$

With a $1000 monthly payment, the present value of their loan would be

$$P = 1000\left[\frac{1-\left(1+\frac{0.095}{12}\right)^{-360}}{\frac{0.095}{12}}\right] \approx 118,926.68, \text{ or } \$118,926.68$$

Since they intend to make a $25,000 down payment, the range of homes they should consider is $120,141 to $143,927.

25. The lower limit of their investment is

$$A = 800\left[\frac{1-\left(1+\frac{0.09}{12}\right)^{-180}}{\frac{0.09}{12}}\right] + 25,000 \approx 103,874.73$$

or approximately $103,875. The upper limit of their investment is

$$A = 1000 \left[\frac{1 - \left(1 + \frac{0.09}{12}\right)^{-180}}{\frac{0.09}{12}} \right] + 25{,}000 \approx 123{,}593.41$$

or approximately \$123,593. Therefore, the price range of houses they should consider is \$103,875 to \$123,593.

27. True. See the formula in the text.

USING TECHNOLOGY EXERCISES 5.2, page 309

1. \$59,622.15

3. \$8453.59

5. :PROGRAM: PVAN
   :Disp "R"
   :Input R
   :Disp "i"
   :Input i
   :Disp "N"
   :Input N
   :(R/i)(1-(1+i)^(-N))→ P
   :Disp "AMOUNT IS"
   :Disp P

7. \$45,983.53

9. \$18,344.08

EXERCISES 5.3, page 317

1. The size of each installment is given by
$$R = \frac{100{,}000(0.08)}{1 - (1 + 0.08)^{-10}} \approx 14{,}902.95, \text{ or } \$14{,}902.95.$$

3. The size of each installment is given by
$$R = \frac{5000(0.01)}{1 - (1 + 0.01)^{-12}} \approx 444.24, \text{ or } \$444.24.$$

5. The size of each installment is given by $R = \dfrac{25{,}000(0.0075)}{1 - (1 + 0.0075)^{-48}} \approx 622.13, \text{ or } \$622.13.$

7. The size of each installment is
$$R = \frac{80,000(0.00875)}{1-(1+0.00875)^{-360}} \approx 731.79, \text{ or } \$731.79.$$

9. The periodic payment that is required is
$$R = \frac{20,000(0.02)}{(1+0.02)^{12} - 1} \approx 1491.19, \text{ or } \$1491.19.$$

11. The periodic payment that is required is
$$R = \frac{100,000(0.0075)}{(1+0.0075)^{120} - 1} \approx 516.76, \text{ or } \$516.76$$

13. The periodic payment that is required is
$$R = \frac{250,000(0.00875)}{(1+0.00875)^{300} - 1} \approx 172.95, \text{ or } \$172.95.$$

15. The size of each installment is given by
$$R = \frac{100,000(0.10)}{1-(1+0.10)^{-10}} \approx 16,274.54, \text{ or } \$16,274.54.$$

17. The monthly payment in each case is given by
$$R = \frac{100,000\left(\dfrac{r}{12}\right)}{1-\left(1+\dfrac{r}{12}\right)^{-360}}$$

Thus, if $r = 0.08$, then $R = \dfrac{100,000\left(\dfrac{0.08}{12}\right)}{1-\left(1+\dfrac{0.08}{12}\right)^{-360}} \approx 733.76, \text{ or } \$733.76$

If $r = 0.09$, then $R = \dfrac{100,000\left(\dfrac{0.09}{12}\right)}{1-\left(1+\dfrac{0.09}{12}\right)^{-360}} \approx 804.62, \text{ or } \$804.62$

If $r = 0.10$, then $R = \dfrac{100{,}000\left(\dfrac{0.10}{12}\right)}{1-\left(1+\dfrac{0.10}{12}\right)^{-360}} \approx 877.57$, or \$877.57.

If $r = 0.11$, then $R = \dfrac{100{,}000\left(\dfrac{0.11}{12}\right)}{1-\left(1+\dfrac{0.11}{12}\right)^{-360}} \approx \$952.32.$

a. The difference in monthly payments in the two loans is
$877.57 - $665.30 = $212.27.
b. The monthly mortgage payment on a \$150,000 mortgage would be
1.5(\$877.57) = \$1316.36.
The monthly mortgage payment on a \$50,000 mortgage would be
0.5(\$877.57) = \$438.79.

19. a. The amount of the loan required is 16000 - (0.25)(12000) or 13,000 dollars. If the car is financed over 36 months, the payment will be

$$R = \dfrac{12{,}000\left(\dfrac{0.10}{12}\right)}{1-\left(1+\dfrac{0.10}{12}\right)^{-36}} \approx 387.21, \text{ or } \$387.21 \text{ per month.}$$

If the car is financed over 48 months, the payment will be

$$R = \dfrac{12{,}000\left(\dfrac{0.10}{12}\right)}{1-\left(1+\dfrac{0.10}{12}\right)^{-48}} \approx 304.35, \text{ or } \$304.35 \text{ per month.}$$

b. The interest charges for the 36-month plan are
36(387.21) - 12000 = 1939.56,
or \$1939.56. The interest charges for the 48-month plan are
48(304.35) - 12000 = 2608.80, or \$2608.80.

21. The amount borrowed is $180,000 - 20,000 = 160,000$ dollars. The size of the monthly installment is

$$R = \frac{160,000\left(\dfrac{0.08}{12}\right)}{1-\left(1+\dfrac{0.08}{12}\right)^{-360}} \approx 1174.0233, \text{ or } \$1174.02.$$

To find their equity after five years, we compute

$$P = 1174.02\left[\frac{1-\left(1+\dfrac{0.08}{12}\right)^{-300}}{\dfrac{0.08}{12}}\right] \approx 152,112$$

or $152,112, and so their equity is

$$180,000 - 152,112 = 27,888, \text{ or } \$27,888.$$

To find their equity after ten years, we compute

$$P = 1174.02\left[\frac{1-\left(1+\dfrac{0.08}{12}\right)^{-240}}{\dfrac{0.08}{12}}\right] \approx 140,359, \text{ or } \$140,359.$$

and their equity is $180,000 - 140,359 = 39,641$, or $39,641.
To find their equity after twenty years, we compute

$$P = 1174.02\left[\frac{1-\left(1+\dfrac{0.08}{12}\right)^{-120}}{\dfrac{0.08}{12}}\right] \approx 96,764, \text{ or } \$96,764,$$

and their equity is $180,000 - 96,764$, or $83,236.

23. The amount that must be deposited quarterly into this fund is

$$R = \frac{\left(\dfrac{0.09}{4}\right)200,000}{\left(1+\dfrac{0.09}{4}\right)^{40}-1} \approx 3,135.48, \text{ or } \$3,135.48.$$

5 Mathematics of Finance

25. The size of each quarterly installment is given by

$$R = \frac{\left(\dfrac{0.10}{4}\right)20{,}000}{\left(1+\dfrac{0.10}{4}\right)^{12} - 1} \approx 1449.74, \text{ or } \$1449.74.$$

27. The value of the IRA account after 20 years is

$$S = 375\left[\frac{\left(1+\dfrac{0.08}{4}\right)^{80} - 1}{\dfrac{0.08}{4}}\right] \approx 72{,}664.48, \text{ or } \$72{,}664.48.$$

The payment he would receive at the end of each quarter for the next 15 years is given by

$$R = \frac{\left(\dfrac{0.08}{4}\right)72{,}664.48}{1-\left(1+\dfrac{0.08}{4}\right)^{-60}} \approx 2090.41, \text{ or } \$2090.41.$$

If he continues working and makes quarterly payments until age 65, the value of the IRA account would be

$$S = 375\left[\frac{\left(1+\dfrac{0.08}{4}\right)^{100} - 1}{\dfrac{0.08}{4}}\right] \approx 117{,}087.11, \text{ or } \$117{,}087.11.$$

The payment he would receive at the end of each quarter for the next 10 years is given by

$$R = \frac{\left(\dfrac{0.08}{4}\right)117{,}087.11}{1-\left(1+\dfrac{0.08}{4}\right)^{-40}} \approx 4280.21, \text{ or } \$4280.21.$$

29. The monthly payment the Sandersons are required to make under the terms of their original loan is given by

$$R = \frac{100,000\left(\dfrac{0.10}{12}\right)}{1-\left(1+\dfrac{0.10}{12}\right)^{-240}} \approx 965.02, \text{ or } \$965.02.$$

The monthly payment the Sandersons are required to make under the terms of their new loan is given by

$$R = \frac{100,000\left(\dfrac{0.078}{12}\right)}{1-\left(1+\dfrac{0.078}{12}\right)^{-240}} \approx 824.04, \text{ or } \$824.04.$$

The amount of money that the Sandersons can expect to save over the life of the loan by refinancing is given by
$$240(965.02 - 824.04) = 33,835.20, \text{ or } \$33,835.20.$$

31. a. The monthly payment required to amortize the loan over the life of the loan under option A is given by

$$\frac{150,000\left(\dfrac{0.075}{12}\right)}{1-\left(1+\dfrac{0.075}{12}\right)^{-360}} = 1048.82, \text{ or } \$1048.82/\text{month}.$$

The monthly payment required to amortize the loan over the life of the loan under option B is given by

$$\frac{150,000\left(\dfrac{0.075}{12}\right)}{1-\left(1+\dfrac{0.075}{12}\right)^{-180}} = 1369.29, \text{ or } \$1369.29.$$

b. The interest paid under option A is given by
$$(1048.82)(360)-150,000 = 227,575.20, \text{ or } \$227,575.20.$$

Similarly, the interest paid under option B is given by
$$(1369.29)(180) - 150,000 = 96,472.20, \text{ or } \$96,472.20.$$

Then the difference between these two amounts gives the amount saved by Mr. and Mrs. Martinez if they chose the 15-year mortgage instead of the 30-year mortgage. Thus,
$$227,575.20 - 96,472.20 = 131,103, \text{ or } \$131,103$$
is the amount that would be saved over the life of the loan.

1. $3645.40   3. $18,443.75

5. :PROGRAM: SINKFD   7. $916.26   9. $809.31
   :Disp "S"
   :Input S
   :Disp "i"
   :Input i
   :Disp "N"
   :Input N
   :S*i/((1+i)^N-1)→R
   :Disp "R is"
   :Disp R

11. $45,069.31. The amortization schedule follows.

End of Period	Interest charged	Repayment made	Payment toward Principal	Outstanding Principal
0				265,000.00
1	19,610.00	45,069.31	25,459.31	239,540.69
2	17,726.01	45,069.31	27,343.30	212,197.39
3	15,702.61	45,069.31	29,366.70	182,830.69
4	13,529.47	45,069.31	31,539.84	151,290.85
5	11,195.52	45,069.31	33,873.79	117,417.06
6	8,688.86	45,069.31	36,380.45	81,036.61
7	5,996.71	45,069.31	39,072.60	41,964.01
8	3,105.34	45,069.35	41,964.01	.00

EXERCISES 5.4, page 328

1. $a_9 = 6 + (9 - 1)3 = 30$

3. $a_8 = -15 + (8 - 1)\left(\dfrac{3}{2}\right) = -\dfrac{9}{2} = -4.5.$

5. $a_{11} - a_4 = (a_1 + 10d) - (a_1 + 3d) = 7d.$ Also, $a_{11} - a_4 = 107 - 30 = 77.$
   Therefore, $7d = 77$, and $d = 11$. Next,
   $$a_4 = a + 3d = a + 3(11) = a + 33 = 30.$$
   and $a = -3$. Therefore, the first five terms are -3, 8, 19, 30, 41.

7. Here $a = x$, $n = 7$, and $d = y$. Therefore, the required term is
   $$a_7 = x + (7 - 1)y = x + 6y.$$

9. Using the formula for the sum of the terms of an arithmetic progression with $a = 4$, $d = 7$ and $n = 15$, we have
   $$S_n = \frac{n}{2}[2a + (n-1)d]]$$
   $$S_{15} = \frac{15}{2}[2(4) + (15-1)7] = \frac{15}{2}(106) = 795$$

11. The common difference is $d = 2$ and the first term is $a = 15$. Using the formula for the nth term $a_n = a + (n - 1)d$,

    we have $57 = 15 + (n - 1)(2) = 13 + 2n$
    $$2n = 44, \quad \text{and} \quad n = 22.$$
    Using the formula for the sum of the terms of an arithmetic progression with $a = 15$, $d = 2$ and $n = 22$, we have
    $$S_n = \frac{n}{2}[2a + (n-1)d]]$$
    $$S_{22} = \frac{22}{2}[2(15) + (22-1)2] = 11(72) = 792.$$

13. $$f(1) + f(2) + f(3) + \cdots + f(20)$$
    $$= [3(1) - 4] + [3(2) - 4] + [3(3) - 4] + \cdots + [3(20) - 4]$$
    $$= 3(1 + 2 + 3 + \cdots + 20) + 20(-4)$$
    $$= 3\left(\frac{20}{2}\right)[2(1) + (20-1)1] - 80$$
    $$= 550.$$

15. $$S_n = \frac{n}{2}[2a_1 + (n-1)d]] = \frac{n}{2}(a_1 + a_1 + (n-1)d]$$

$$= \frac{n}{2}(a_1 + a_n)$$

a. $S_{11} = \frac{11}{2}(3 + 47) = 275.$

b. $S_{20} = \frac{20}{2}[5 + (-33)] = -280.$

17. Let n be the number of weeks till she reaches 10 miles. Then
$$a_n = 1 + (n-1)\frac{1}{4} = 1 + \frac{1}{4}n - \frac{1}{4} = \frac{1}{4}n + \frac{3}{4} = 10$$
Therefore, $n + 3 = 40$, and $n = 37$; that is, it will take Karen 37 weeks to meet her goal.

19. To compute the tourist's fare by taxi, take $a = 1$, $d = 0.60$, and $n = 25$. Then the required fare is given by
$$a_{25} = 1 + (25 - 1)0.60 = 15.4,$$
or $15.40. Therefore, by taking the airport limousine, the tourist will save
15.40 - 7.50 = 7.90, or $7.90.

21. a. Using the formula for the sum of an arithmetic progression, we have
$$S_n = \frac{n}{2}[2a + (n-1)d]]$$
$$= \frac{N}{2}[2(1) + (N-1)(1)]$$
$$= \frac{N}{2}(N+1).$$

b. $S_{10} = \frac{10}{2}(10 + 1) = 5(11) = 55$

$$D_3 = (C - S)\frac{N - (n-1)}{55} = (6000 - 500)\frac{10 - (3-1)}{55} = 5500\left(\frac{8}{55}\right)$$
$$= 800, \text{ or } \$800.$$

23. This is a geometric progression with $a = 4$ and $r = 2$. Next, $a_7 = 4(2)^6 = 256$,

and $\qquad S_7 = \frac{4(1 - 2^7)}{1 - 2} = 508.$

25. If we compute the ratios

$$\frac{a_2}{a_1} = \frac{-\frac{3}{8}}{\frac{1}{2}} = -\frac{3}{4} \quad \text{and} \quad \frac{a_3}{a_2} = \frac{\frac{1}{4}}{-\frac{3}{8}} = -\frac{2}{3},$$

we see that the given sequence is not geometric since the ratios are not equal.

27. This is a geometric progression with $a = 243$, and $r = 1/3$.

$$a_7 = 243\left(\tfrac{1}{3}\right)^6 = \tfrac{1}{3}.$$

$$S_7 = \frac{243\left(1 - \left(\tfrac{1}{3}\right)^7\right)}{1 - \tfrac{1}{3}} = 364\tfrac{1}{3}.$$

29. First, we compute

$$r = \frac{a_2}{a_1} = \frac{3}{-3} = -1.$$

Next, $a_{20} = -3(-1)^{19} = 3$ and so $S_{20} = \dfrac{-3[1 - (-1)^{20}]}{[1 - (-1)]} = 0$.

31. The population in five years is expected to be
$$200{,}000(1.08)^{6-1} = 200{,}000(1.08)^5 \approx 293{,}866.$$

33. The salary of a union member whose salary was \$22,000 six years ago is given by the 7th term of a geometric progression whose first term is 22,000 and whose common ratio is 1.11. Thus
$$a_7 = (22{,}000)(1.11)^6 = 41{,}149.12, \text{ or } \$41{,}149.12.$$

35. With 8 percent raises per year, the employee would make
$$S_4 = 28{,}000\left[\frac{1 - (1.08)^4}{1 - 1.08}\right] \approx 126{,}171.14,$$

or \$126,171.14 over the next four years.
With \$1500 raises per year, the employee would make

$$S_4 = \frac{4}{2}[2(28{,}000) + (4 - 1)1500] = 121{,}000$$

or \$121,000 over the next four years. We conclude that the employee should choose the 8 percent per year raises.

37. *a.* During the sixth year, she will receive
$$a_6 = 10{,}000(1.15)^5 \approx 20{,}113.57, \text{ or } \$20{,}113.57.$$

b. The total amount of the six payments will be given by

$$S_6 = \frac{10,000[1-(1.15)^6]}{1-1.15} \approx 87,537.38, \text{ or } \$87,537.38.$$

39. The book value of the office equipment at the end of the eighth year is given by

$$V(8) = 150,000\left(1-\frac{2}{10}\right)^8 \approx 25,165.82, \text{ or } \$25,165.82.$$

41. The book value of the restaurant equipment at the end of six years is given by
$$V(6) = 150,000(0.8)^6 \approx 39,321.60,$$
or $39,321.60. By the end of the sixth year, the equipment will have depreciated by
$$D(n) = 150,000 - 39,321.60 = 110,678.40, \text{ or } \$110,678.40.$$

43. True. Suppose d is the common difference of $a_1, a_2, ..., a_n$ and e is the common difference of $b_1, b_2, ..., b_n$. Then $d+e$ is the common difference of $a_1+b_1, a_2+b_2, ..., a_n+b_n$ is obtained by adding the constant $c+d$ to it, we see that it is indeed an arithmetic progression.

CHAPTER 5, REVIEW EXERCISES, page 332

1. a. Here $P = 5000$, $r = 0.1$, and $m = 1$. Thus, $i = r = 0.1$ and $n = 4$. So
$A = 5000(1.1)^4 = 7320.5$, or $7320.50.

    b. Here $m = 2$ so that $i = 0.1/2 = 0.05$ and $n = (4)(2) = 8$. So
$A = 5000(1.05)^8 \approx 7387.28$ or $7387.28.

    c. Here $m = 4$, so that $i = 0.1/4 = 0.025$ and $n = (4)(4) = 16$. So
$A = 5000(1.025)^{16} \approx 7,422.53$, or $7422.53.

    d. Here $m = 12$, so that $i = 0.1/12$ and $n = (4)(12) = 48$. So
$$A = 5000\left(1+\frac{0.10}{12}\right)^{48} \approx 7446.77, \text{ or } \$7446.77.$$

3. a. The effective rate of interest is given by
$$r_{eff} = \left(1+\frac{r}{m}\right)^m - 1 = (1+0.12) - 1 = 0.12, \text{ or } 12 \text{ percent.}$$

    b. The effective rate of interest is given by

$$r_{eff} = \left(1 + \frac{r}{m}\right)^m - 1 = \left(1 + \frac{0.12}{2}\right)^2 - 1 = 0.1236, \text{ or } 12.36 \text{ percent.}$$

c. The effective rate of interest is given by

$$r_{eff} = \left(1 + \frac{r}{m}\right)^m - 1 = \left(1 + \frac{0.12}{4}\right)^4 - 1 \approx 0.125509, \text{ or } 12.5509 \text{ percent.}$$

d. The effective rate of interest is given by

$$r_{eff} = \left(1 + \frac{r}{m}\right)^m - 1 = \left(1 + \frac{0.12}{12}\right)^{12} - 1 \approx 0.126825, \text{ or } 12.6825 \text{ percent.}$$

5. The present value is given by

$$P = 41,413\left(1 + \frac{0.065}{4}\right)^{-20} \approx 30,000.29, \text{ or approximately } \$30,000.$$

7.

$$S = 150\left[\frac{\left(1 + \frac{0.08}{4}\right)^{28} - 1}{\frac{0.08}{4}}\right] \approx 5557.68, \text{ or } \$5557.68.$$

9. Using the formula for the present value of an annuity with $R = 250$, $n = 36$, $i = 0.09/12 = 0.0075$, we have

$$P = 250\left[\frac{1 - (1.0075)^{-36}}{0.0075}\right] \approx 7861.70, \text{ or } \$7861.70.$$

11. Using the amortization formula with $P = 22,000$, $n = 36$, and $i = 0.085/12$, we find

$$R = \frac{22,000\left(\frac{0.085}{12}\right)}{1 - \left(1 + \frac{0.085}{12}\right)^{-36}} \approx 694.49, \text{ or } \$694.49.$$

13. Using the sinking fund formula with $S = 18,000$, $n = 48$, and $i = 0.06/12$, we have

$$R = \frac{\left(\dfrac{0.06}{12}\right)18{,}000}{\left(1+\dfrac{0.06}{12}\right)^{48}-1} \approx 332.73, \text{ or } \$332.73.$$

15. We are asked to find r such that

$$\left(1+\frac{r}{365}\right)^{365} = \left(1+\frac{0.072}{12}\right)^{12} = 1.074424168.$$

Then
$$1+\frac{r}{365} = (1.074424168)^{1/365}$$

$$\frac{r}{365} = (1.074424168)^{1/365} - 1$$

and
$$r = 365[1.074424168^{1/365} - 1]$$

$$\approx 0.071791919, \text{ or approximately } 7.179 \text{ percent.}$$

17. Let a_n denote the sales during the nth year of operation. Then

$$\frac{a_{n+1}}{a_n} = 1.14 = r.$$

Therefore, the sales during the fourth year of operation were
$$a_4 = a_1 r^{n-1} = 1{,}750{,}000(1.14)^{4-1} = 2{,}592{,}702, \text{ or } \$2{,}592{,}702.$$

The total sales over the first four years of operation are given by
$$S_4 = \frac{a(1-r^4)}{1-r} = \frac{1{,}750{,}000[1-(1.14)^4]}{1-(1.14)} = 8{,}612{,}002,$$

or \$8,612,002.

19. Using the present value formula for compound interest, we have

$$P = A\left(1+\frac{r}{m}\right)^{-mt} = 19{,}440.31\left(1+\frac{0.065}{12}\right)^{-12(4)} = 15{,}000.00, \text{ or } \$15{,}000.$$

21. Using the sinking fund formula with $S = 40{,}000$, $n = 120$, and $i = 0.08/12$, we find

$$R = \frac{\left(\dfrac{0.08}{12}\right)40,000}{\left(1+\dfrac{0.08}{12}\right)^{120}-1} = 218.64, \text{ or } \$218.64.$$

23. Using the formula for the present value of an annuity, we see that the equivalent cash payment of Maria's auto lease is

$$P = 300\frac{1-\left(1+\dfrac{0.05}{12}\right)^{-48}}{\dfrac{0.05}{12}} = 13,026.89, \text{ or } \$13,026.89.$$

25. a. The monthly payment is given by
$$P = \frac{(120,000)(0.0075)}{1-(1+0.0075)^{-360}} \approx 965.55, \text{ or } \$965.55.$$

b. We can find the total interest payment by computing
$$360(965.55) - 120,000 \approx 227,598, \text{ or } \$227,598.$$

c. We first compute the present value of their remaining payments. Thus,

$$P = 965.55\left[\frac{1-(1+0.0075)^{-240}}{0.0075}\right] \approx 107,316.01.$$

or $107,316.01. Then their equity is 150,000 - 107,316.01, or approximately $42,684.

27. Using the sinking fund formula with $S = 500,000$, $n = 20$, and $i = 0.10/4$, we find that the amount of each installment should be

$$R = \frac{\left(\dfrac{0.10}{4}\right)500,000}{\left(1+\dfrac{0.10}{4}\right)^{20}-1} \approx 19,573.56, \text{ or } \$19,573.56.$$

29. Using the amortization formula, we find that Bill's monthly payment will be

$$R = \frac{3200\left(\dfrac{0.186}{12}\right)}{1-\left(1+\dfrac{0.186}{12}\right)^{-18}} \approx 205.09, \text{ or } \$205.09.$$

CHAPTER 6

EXERCISES 6.1, page 344

1. $\{x \mid x$ is gold medalist in the 2000 Summer Olympic Games$\}$

3. $\{x \mid x$ is an integer greater than 2 and less than 8$\}$

5. $\{2,3,4,5,6\}$

7. $\{-2\}$

9. a. True--the order in which the elements are listed is not important.
   b. False-- A is a set, not an element.

11. a. False. The empty set has no elements.
    b. False. 0 is an element and $\varnothing$ is a set.

13. True.

15. a. True. 2 belongs to A.
    b. False. For example, 5 belongs to A but $5 \notin \{2,4,6\}$.

17. a. and b.

19. a. $\varnothing, \{1\}, \{2\}, \{1,2\}$

    b. $\varnothing, \{1\}, \{2\}, \{3\}, \{1,2\}, \{1,3\}, \{2,3\}, \{1,2,3\}$

    c. $\varnothing, \{1\}, \{2\}, \{3\}, \{4\}, \{1,2\}, \{1,3\}, \{1,4\}, \{2,3\}, \{2,4\}, \{3,4\}, \{1,2,3\},$
       $\{1,2,4\}, \{2,3,4\}, \{1,3,4\}, \{1,2,3,4\}$

21. $\{1, 2, 3, 4, 6, 8, 10\}$

23. $\{$Jill, John, Jack, Susan, Sharon$\}$

25. a.

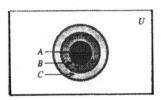

b.

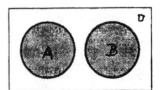

c.

27. a.

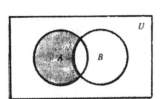

b.

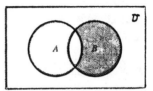

29. a.

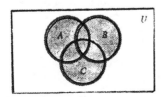

b.

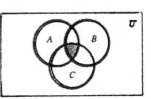

31. a.

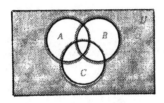

b.

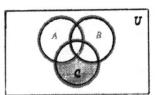

33. a. $A^C = \{2,4,6,8,10\}$
    b. $B \cup C = \{2,4,6,8,10\} \cup \{1,2,4,5,8,9\} = \{1,2,4,5,6,8,9,10\}$
    c. $C \cup C' = U = \{1,2,3,4,5,6,7,8,9,10\}$

35. a. $(A \cap B) \cup C = C = \{1,2,4,5,8,9\}$
    b. $(A \cup B \cup C)^c = \varnothing$
    c. $(A \cap B \cap C)^c = U = \{1,2,3,4,5,6,7,8,9,10\}$

37. a. The sets are not disjoint. 4 is an element of both sets.
    b. The sets are disjoint as they have no common elements.

39. a. The set of all employees at the Universal Life Insurance Company who do not drink tea.
    b. The set of all employees at the Universal Life Insurance Company who do not drink coffee.

41. a. The set of all employees at the Universal Life Insurance Company who drink tea but not coffee.
    b. The set of all employees at the Universal Life Insurance Company who drink coffee but not tea.

43. a. The set of all employees at the hospital who are not doctors
    b. The set of all employees at the hospital who are not nurses

45. a. The set of all employees at the hospital who are female doctors.
    b. The set of all employees at the hospital who are both doctors and administrators.

47. a. $D \cap F$ b. $R \cap F^c \cap L^c$

49. a. B^C b. $A \cap B$ c. $A \cap B \cap C^c$

51. a. Region 1: $A \cap B \cap C$ is the set of tourists who used all three modes of transportation over a 1-week period in London.
    b. Regions 1 and 4: $A \cap C$ is the set of tourists who have taken the underground and a bus over a 1-week period in London.
    c. Regions 4, 5, 7, and 8: B^c is the set of tourists who have not taken a cab over a 1-week period in London.

53. $A \subset A \cup B$ $B \subset A \cup B$

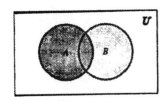   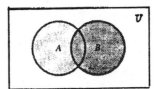

55. $A \cup (B \cup C) = (A \cup B) \cup C$

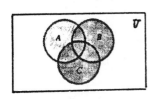   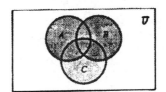

57. $A \cap (B \cup C) = (A \cap B) \cup (A \cap C)$

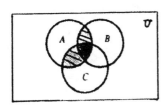 $=$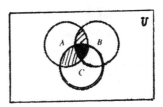

59. a. $A \cup (B \cup C) = \{1,3,5,7,9\} \cup (\{1,2,4,7,8\} \cup \{2,4,6,8\})$
$\qquad\qquad = \{1\ 3,5,7,9\} \cup \{1,2,4,6,7,8\}$
$\qquad\qquad = \{1,2,3,4,5,6,7,8,9\}$

$\quad (A \cup B) \cup C = (\{1,3,5,7,9\} \cup (\{1,2,4,7,8\}) \cup \{2,4,6,8\})$
$\qquad\qquad\quad = \{1,2,3,4,5,7,8,9\} \cup \{2,4,6,8\}$
$\qquad\qquad\quad = \{1,2,3,4,5,6,7,8,9\}$

b. $A \cap (B \cap C) = \{1,3,5,7,9\} \cap (\{1,2,4,7,8\} \cap \{2,4,6,8\})$
$\qquad\qquad\quad = \{1,3,5,7,9\} \cap (\{2,4,8\}$
$\qquad\qquad\quad = \varnothing$

$$(A \cap B) \cap C = (\{1,3,5,7,9\} \cap \{1,2,4,7,8\}) \cap \{2,4,6,8\}$$
$$= \{1,7\} \cap \{2,4,6,8\}$$
$$= \varnothing.$$

61. a. r, u, v, w, x, y b. v, r

63. a. t, y, s b. t, s, w, x, z

65. $A \subset C$

67. False. Since every element in a set A belongs to A, A is a subset of itself.

69. True. If at least one of the sets A or B is nonempty, then $A \cup B \ne \varnothing$.

71. True. $(A \cup A^c)^c = U^c = \varnothing$.

EXERCISES 6.2, page 352

1. $A \cup B = \{a,e,g,h,i,k,l,m,o,u\}$, and so $n(A \cup B) = 10$. Next,
    $n(A) + n(B) = 5 + 5 = 10$.

3. a. $A = \{2,4,6,8\}$ and $n(A) = 4$.
    b. $B = \{6,7,8,9,10\}$ and $n(B) = 5$
    c. $A \cup B = \{2,4,6,7,8,9,10\}$ and $n(A \cup B) = 7$.
    d. $A \cap B = \{6,8\}$ and $n(A \cap B) = 2$.

5. $A \cup B = \{a,e,i,o,u\} \cup \{b,d,e,o,u\} = \{a,b,d,e,i,o,u\}$ and $n(A \cup B) = 7$
    $A = \{a,e,i,o,u\}$ so $n(A) = 5$, $B = \{b,d,e,o,u\}$ and $n(B) = 5$, and
    $A \cap B = \{a,e,i,o,u\} \cap \{b,d,e,o,u\} = \{e,o,u\}$ so that $n(A \cap B) = 3$.
    Therefore, $n(A \cup B) = n(A) + n(B) - n(A \cap B) = 5 + 5 - 3 = 7$.

7. $n(A \cap B) = n(A) + n(B) - n(A \cup B) = 10 + 8 - 15 = 3$.

9. Refer to the Venn diagram at the right.
    a. $n(A^C \cap B) = 40$.
    b. $n(B^C) = 60 + 60 = 120$.
    c. $n(A^C \cap B^C) = n(A \cup B)^C = 60$.

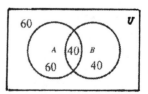

11. $n(A \cup B) = n(A) + n(B) - n(A \cap B)$ so
$$n(A) = n(A \cup B) + n(A \cap B) - n(B)$$
$$= 14 + 3 - 6 = 11.$$

13. $n(A \cap B \cap C)$
$$= n(A \cup B \cup C) - n(A) - n(B) - n(C) + n(A \cap B) + n(A \cap C) + n(B \cap C)$$
$$= 31 - 16 - 16 - 14 + 6 + 5 + 6 = 2.$$

15. Let $A = \{x \mid x$ is a subscriber to the daily morning edition$\}$ and $B = \{x \mid x$ is a subscriber to the Sunday *L.A. Times*$\}$. Then, we are given that $n(A) = 900$, $n(A \cap B) = 500$, and $n(A \cup B) = 1000$. Refer to the Venn diagram at the right. Since
$$n(A \cup B) = n(A) + n(B) - n(A \cap B),$$
we see that
$$n(B) = n(A \cup B) + n(A \cap B) - n(A)$$
$$= 1000 + 500 - 900 = 600$$
Next, $(B \cap A^c) = n(B) - n(A \cap B) = 600 - 500 = 100$.

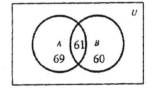

17. Let A denote the set of prisoners in the Wilton County Jail who were accused of a felony and B the set of prisoners in that jail who were accused of a misdemeanor. Then we are given that
$$n(A \cup B) = 190$$
Refer to the diagram at the right.

Then the number of prisoners who were accused of both a felony and a misdemeanor is given by $(A \cap B) = n(A) + n(B) - n(A \cup B) = 130 + 121 - 190 = 61$.

19. Let U denote the set of all customers surveyed, and let
$$A = \{x \in U \mid x \text{ buys brand } A\}$$
$$B = \{x \in U \mid x \text{ buys brand } B\}.$$
Then $n(U) = 120$, $n(A) = 80$, $n(B) = 68$, and $n(A \cap B) = 42$.

Refer to the diagram at the right.

a. The number of customers who buy at least
   one of these brands is

   $n(A \cup B) = 80 + 68 - 42 = 106.$

b. The number who buy exactly one of these brands is

   $n(A \cap B^c) + n(A^c \cap B) = 38 + 26 = 64$

c. The number who buy only brand A is

   $n(A \cap B^c) = 38.$

d. The number who buy none of these brands is

   $n[(A \cup B)^c] = 120 - 106 = 14.$

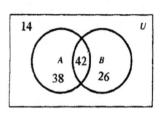

21. Let U denote the set of 200 investors and let

   $A = \{ x \in U \mid x \text{ uses a discount broker}\}$

   $B = \{ x \in U \mid x \text{ uses a full-service broker}\}.$

   Refer to the diagram at the right.

   a. The number of investors who use at least one
   kind of broker is

   $n(A \cup B) = n(A) + n(B) - n(A \cap B)$

   $\qquad\qquad = 120 + 126 - 64 = 182.$

   b. The number of investors who use exactly one kind of broker is

   $n(A \cap B^c) + n(A^c \cap B) = 56 + 62 = 118.$

   c. The number of investors who use only discount brokers is

   $n(A \cap B^c) = 56.$

   d. The number of investors who don't use a broker is

   $n(A \cup B)^c = n(U) - n(A \cup B) = 200 - 182 = 18.$

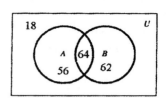

In Exercises 23 and 25, refer to the figure that follows.

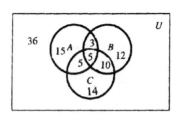

23. a. $n[A \cap (B \cup C)] = 13$ b. $n[A \cap (B \cup C)^c] = 15$

25. a. $n[A \cup (B \cap C)] = 38$

b. $n(A^c \cap B^c \cap C^c)^c = n[(A \cup B \cup C)] = 64.$

27. Let U denote the set of all economists surveyed, and let
$A = \{ x \in U | x$ had lowered his estimate of the consumer inflation rate$\}$
$B = \{ x \in U | x$ had raised his estimate of the GDP growth rate$\}$.
Refer to the diagram at the right.
Then $n(U) = 10$, $n(A) = 7$, $n(B) = 8$,
and $n(A \cap B^c) = 2$. Then the number
of economists who had both lowered
their estimate of the consumer inflation
rate and raised their estimate of the GDP
rate is given by
$n(A \cap B) = 5.$

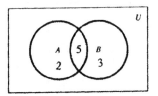

29. Let U denote the set of 100 college students who were surveyed and let
$A = \{ x \in U | x$ is a student who reads *Time* magazine$\}$
$B = \{ x \in U | x$ is a student who reads *Newsweek* magazine$\}$
and
$C = \{ x \in U | x$ is a student who reads *U.S. News and World Report* magazine$\}$
Then $n(A) = 40$, $n(B) = 30$, $n(C) = 25$, $n(A \cap B) = 15$,
$n(A \cap C) = 12$, $n(B \cap C) = 10$, and $n(A \cap B \cap C) = 4.$

Refer to the diagram on the right.

a. The number of students surveyed who read
at least one magazine is
$n(A \cup B \cup C) = 17 + 11 + 4 + 8 + 6 + 7 + 9 = 62$

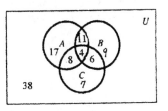

b. The number of students surveyed who read exactly one magazine is
$n(A \cap B^c \cap C^c) + n(A^c \cap B \cap C^c) + n(A^c \cap B^c \cap C)$
$= 17 + 9 + 7 = 33.$

c. The number of students surveyed who read exactly two magazines is
$$n(A \cap B \cap C^c) + n(A^c \cap B \cap C) + n(A \cap B^c \cap C)$$
$$= 11 + 6 + 8 = 25.$$

d. The number of students surveyed who did not read any of these magazines is
$$n(A \cup B \cup C)^c = 100 - 62 = 38.$$

31. Let U denote the set of all customers surveyed, and let

$A = \{ x \in U \mid x \text{ buys brand } A \}$

$B = \{ x \in U \mid x \text{ buys brand } B \}.$

$C = \{ x \in U \mid x \text{ buys brand } C \}.$

Refer to the figure at the right.
Then

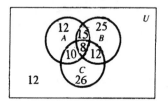

$n(U) = 120, \ n(A \cap B \cap C^c) = 15,$
$\quad n(A^c \cap B \cap C^c) = 25, \ n(A^c \cap B^c \cap C) = 26,$
$\quad n(A \cap B \cap C^c) = 15, \ n(A \cap B^c \cap C) = 10,$
$\quad n(A^c \cap B \cap C) = 12, \text{ and } n(A \cap B \cap C) = 8.$

a. The number of customers who buy at least one of these brands is
$$n(A \cup B \cup C) = 12 + 15 + 25 + 12 + 8 + 10 + 26 = 108.$$
b. The number who buy labels A and B but not C is
$$n(A \cap B \cap C^c) = 15$$
c. The number who buy brand A is
$$n(A) = 12 + 10 + 15 + 8 = 45.$$
d. The number who buy none of these brands is
$$n[(A \cup B \cup C)^c] = 120 - 108 = 12.$$

33. True. If $A \subseteq B$, then $B = A \cup (A^c \cap B)$ and $A \cap (A^c \cap B) = \emptyset$. Therefore, $n(B) = n(A) + n(A^c \cap B)$.

35. Write Equation (4) as $n(D \cup E) = n(D) + n(E) - n(D \cap E)$. Then let $D = A \cup B$ and $E = C$ so that

$$n(A \cup B \cup C) = n(A \cup B) + n(C) - n[(A \cup B) \cap C]$$
$$= n(A) + n(B) - n(A \cap B) + n(C) - n[(A \cup B) \cap C]$$
$$= n(A) + n(B) - n(A \cap B) + n(C)$$
$$- \{n[(A \cap C) \cup (B \cap C)]\}$$
$$= n(A) + n(B) - n(A \cap B) + n(C)$$
$$- [n(A \cap C) + n(B \cap C) - n(A \cap C \cap B \cap C)]$$
$$= n(A) + n(B) + n(C) - n(A \cap B)$$
$$- n(A \cap C) - n(B \cap C) + n(A \cap B \cap C)$$

EXERCISES 6.3, page 359

1. By the multiplication principle, the number of rates is given by $(4)(3) = 12$.

3. By the multiplication principle, the number of ways that a blackjack hand can be dealt is $(4)(16) = 64$.

5. By the multiplication principle, she can create $(2)(4)(3) = 24$ different ensembles.

7. The number of paths is $2 \times 4 \times 3$, or 24.

9. By the multiplication principle, we see that the number of ways a health-care plan can be selected is $(10)(3)(2) = 60$.

11. The number of possible social Security numbers is $10^9 = 1,000,000,000$.

13. The number of different responses is
$$\underbrace{(5)\,(5) \ldots (5)}_{50 \text{ terms}} = 5^{50}.$$

15. The number of selections is given by $(2)(5)(3)$, or 30 selections.

17. The number of different selections is $(10)(10)(10)(10) - 10 = 10000 - 10 = 9990$.

19. a. The number of license plate numbers that may be formed is
$$(26)(26)(26)(10)(10)(10) = 17,576,000.$$

   b. The number of license plate numbers that may be formied is
$$(10)(10)(10)(26)(26)(26) = 17,576,000.$$

21. If every question is answered, there are 2^{10}, or 1024, ways. In the second case, there are 3 ways to answer each question, and so we have 3^{10}, or 59,049, ways.

23. The number of ways the first, second and third prizes can be awarded is
$$(15)(14)(13) = 2730.$$

25. The number of ways in which the nine symbols on the wheels can appear in the window slot is (9)(9)(9), or 729. The number of ways in which the eight symbols other than the "lucky dollar" can appear in the window slot is (8)(8)(8) or 512. Therefore, the number of ways in which the "lucky dollars" can appear in the window slot is 729 - 512, or 217.

27. False. Use the Multiplication Principle to conclude that there are $2 \cdot 2 \cdot 2 \cdot 2 \cdot 2 \cdot 2$ or 64 different pizzas.

EXERCISES 6.4, page 372

1. $3(5!) = 3(5)(4)(3)(2)(1) = 360.$

3. $\dfrac{5!}{2!3!} = 5(2) = 10.$

5. $P(5,5) = \dfrac{5!}{(5-5)!} = \dfrac{5!}{0!} = 120$

7. $P(5,2) = \dfrac{5!}{(5-2)!} = \dfrac{5!}{3!} = (5)(4) = 20$

9. $P(n,1) = \dfrac{n!}{(n-1)!} = n$

11. $C(6,6) = \dfrac{6!}{6!0!} = 1$

13. $C(7,4) = \dfrac{7!}{4!3!} = \dfrac{7 \cdot 6 \cdot 5}{3 \cdot 2} = 35$

15. $C(5,0) = \dfrac{5!}{5!0!} = 1$

17. $C(9,6) = \dfrac{9!}{3!6!} = \dfrac{9 \cdot 8 \cdot 7}{3 \cdot 2} = 84$

19. $C(n,2) = \dfrac{n!}{(n-2)!2!} = \dfrac{n(n-1)}{2}$

21. $P(n,n-2) = \dfrac{n!}{(n-(n-2))!} = \dfrac{n!}{(n-n+2)!} = \dfrac{n!}{2}$

23. Order is important here since the word "*glacier*" is different from "*reicalg*", so this is a permutation.

25. Order is not important here. Therefore, we are dealing with a combination. If we

consider a sample of three record-o-phones of which one is defective, it does not matter whether the defective record-o-phone is the first member of our sample, the second member of our sample, or the third member of our sample. The net result is a sample of three record-o-phones of which one is defective.

27. The order is important here. Therefore, we are dealing with a permutation. Consider, for example, 9 books on a library shelf. Each of the 9 books would have a call number, and the books would be placed in order of their call numbers; that is, a call number of 902 would come before a call number of 910.

29. The order is not important here, and consequently we are dealing with a combination. It would not matter if the hand $Q\,Q\,Q\,5\,5$ were dealt or the hand $5\,5\,Q\,Q\,Q$. In each case the hand would consist of three queens and a pair.

31. The number of 4-letter permutations is $P(4,4) = \dfrac{4!}{0!} = 4 \cdot 3 \cdot 2 \cdot 1 = 24$.

33. The number of seating arrangements is $P(4,4) = \dfrac{4!}{0!} = 24$.

35. The number of different batting orders is $P(9,9) = \dfrac{9!}{0!} = 362{,}880$.

37. The number of different ways the 3 candidates can be selected is
$$C(12,3) = \frac{12!}{9!\,3!} = \frac{12 \cdot 11 \cdot 10}{3 \cdot 2 \cdot 1} = 220\,.$$

39. There are 10 letters in the word $ANTARCTICA$, 3As, 1N, 2Ts, 1R, 2Cs, and 1I. Therefore, we use the formula for the permutation of n objects, not all distinct:
$$\frac{n!}{n_1!\,n_2!\,\cdots\,n_r!} = \frac{10!}{3!2!2!} = 151{,}200$$

41. The number of ways the 3 sites can be selected is
$$C(12,3) = \frac{12!}{9!\,3!} = \frac{12 \cdot 11 \cdot 10}{3 \cdot 2 \cdot 1} = 220$$

43. The number of ways in which the sample of 3 transistors can be selected is

$$C(100,3) = \frac{100!}{97!3!} = \frac{100 \cdot 99 \cdot 98}{3 \cdot 2 \cdot 1} = 161,700.$$

45. In this case order is important, as it makes a difference whether a commercial is shown first, last, or in between. The number of ways that the director can schedule the commercials is given by
$$P(6,6) = 6! = 720.$$

47. The inquiries can be directed in
$$P(12,6) = \frac{12!}{6!} = 12 \cdot 11 \cdot 10 \cdot 9 \cdot 8 \cdot 7 = 665,280$$
or 665,280 ways.

49. a. The ten books can be arranged in
$$P(10,10) = 10! = 3,628,800 \text{ ways.}$$

b. If books on the same subject are placed together, then they can be arranged on the shelf
$$P(3,3) \times P(4,4) \times P(3,3) \times P(3,3) = 5184 \text{ ways.}$$
Here we have computed the number of ways the mathematics books can be arranged times the number of ways the social science books can be arranged times the number of ways the biology books can be arranged times the number of ways the 3 sets of books can be arranged.

51. Notice that order is certainly important here.
a. The number of ways that the 20 featured items can be arranged is given by
$$P(20,20) = 20! = 2.43 \times 10^{18}.$$
b. If items from the same department must appear in the same row, then the number of ways they can be arranged on the page is

Number of ways of arranging the rows	x	Number of ways of arranging the items in each of the 5 rows

$$P(5,5) \quad \bullet \quad P(4,4) \times P(4,4) \times P(4,4) \times P(4,4) \times P(4,4)$$
$$= 5! \times (4!)^5 = 955,514,880.$$

53. The number of ways is given by
$$2\{C(2,2) + [C(3,2) - C(2,2)]\} = 2[1 + (3 - 1)] = 2 \times 3 = 6$$

(number of players)[number of ways to win in exactly 2 sets + number of ways to win in exactly 3 sets]

55. The number of ways the measure can be passed is
$$C(3,3) \times [C(8,6) + C(8,7) + C(8,8)] = 37.$$
Here three of the three permanent members must vote for passage of the bill and this can be done in $C(3,3) = 1$ way. Of the 8 nonpermanent members who are voting 6 can vote for passage of the bill, or 7 can vote for passage, or 8 can vote for passage. Therefore, there are
$$C(8,6) + C(8,7) + C(8,8) = 37 \text{ ways}$$
that the nonpermanent members can vote to ensure passage of the measure. This gives $1 \times 37 = 37$ ways that the members can vote so that the bill is passed.

57. a. If no preference is given to any student, then the number of ways of awarding the 3 teaching assistantships is
$$C(12,3) = \frac{12!}{3!9!} = 220.$$
b. If it is stipulated that one particular student receive one of the assistantships, then the remaining two assistantships must be awarded to two of the remaining 11 students. Thus, the number of ways is
$$C(11,2) = \frac{11!}{2!9!} = 55.$$
c. If at least one woman is to be awarded one of the assistantships, and the group of students consists of seven men and five women, then the number of ways the assistantships can be awarded is given by

$$C(5,1) \times C(7,2) + C(5,2) \times C(7,1) + C(5,3)$$
$$= \frac{5!}{4!1!} \cdot \frac{7!}{5!2!} + \frac{5!}{3!2!} \cdot \frac{7!}{6!1!} + \frac{5!}{3!2!} = 105 + 70 + 10 = 185..$$

59. The number of ways of awarding the 7 contracts to 3 different firms is given by
$$P(7,3) = \frac{7!}{4!} = 210.$$
The number of ways of awarding the 7 contracts to 2 different firms is
$$C(7,2) \times P(3,2) = 126. \quad \text{(First pick the two firms, and then award the 7 contracts.)}$$

Therefore, the number of ways the contracts can be awarded if no firm is to receive more than 2 contracts is given by
$$210 + 126 = 336.$$

61. The number of different curricula that are available for the student's consideration is given by

$$C(5,1) \times C(3,1) \times C(6,2) \times C(4,1) + C(5,1) \times C(3,1) \times C(6,2) \times C(3,1)$$

$$= \frac{5!}{4!1!} \cdot \frac{3!}{2!1!} \cdot \frac{6!}{4!2!} \cdot \frac{4!}{3!1!} + \frac{5!}{4!1!} \cdot \frac{3!}{2!1!} \cdot \frac{6!}{4!2!} \cdot \frac{3!}{2!1!}$$

$$= (5)(3)(15)(4) + (5)(3)(15)(3) = 900 + 675 = 1575.$$

63. The number of ways is given by
$$C(10,2) + C(10,1) + C(10,0) = 45 + 10 + 1 = 56$$

or $\quad C(10,8) + C(10,9) + C(10,10) = 56.$

65. The number of ways of dealing a straight flush (5 cards in sequence in the same suit is given by

the number of ways of selecting 5 cards in sequence in the same suit	$\times$	the number of ways of selecting a suit
10	$\bullet$	$C(4,1) = \quad 40.$

67. The number of ways of dealing a flush (5 cards in one suit that are not all in sequence) is given by

the number of ways of selecting 5 cards in one suit	-	the number of ways of selecting 5 cards in one suit in sequence
$4C(13,5)$	-	$4(10)$

$$= 5148 - 40 = 5108.$$

69. The number of ways of dealing a full house (3 of a kind and a pair) is given by

the number of ways of picking 3 of a kind from a given rank	$\times$	the number of ways of picking a pair from the 12 remaining ranks

$$13C(4,3) \bullet 12C(4,2)$$
$$= 13(4) \bullet (12)(6) = 3744.$$

71. The bus will travel a total of 6 blocks. Each route must include 2 blocks running north and south and 4 blocks running east and west. To compute the total number of possible routes, it suffices to compute the number of ways the 2 blocks running north and south can be selected from the six blocks. Thus,

$$C(6,2) = \frac{6!}{2!4!} = 15.$$

73. The number of ways that the quorum can be formed is given by

$$C(12,6) + C(12,7) + C(12,8) + C(12,9) + C(12,10) + C(12,11) + C(12,12)$$
$$= \frac{12!}{6!6!} + \frac{12!}{7!5!} + \frac{12!}{8!4!} + \frac{12!}{9!3!} + \frac{12!}{10!2!} + \frac{12!}{11!1!} + \frac{12!}{12!0!}$$
$$= 924 + 792 + 495 + 220 + 66 + 12 + 1 = 2510.$$

75. Using the formula given in Exercise 74, we see that the number of ways of seating the 5 commentators at a round table is $(5 - 1)! = 4! = 24$.

77. The number of possible corner points is $C(8,3) = \dfrac{8!}{5!3!} = 56$.

79. True.

81. True. $C(n,r) = \dfrac{n!}{(n-r)!r!}$ and $C(n,n-r) = \dfrac{n!}{[n-(n-r)]!(n-r)!} = \dfrac{n!}{r!(n-r)!}$.
So, $C(n,r) = C(n,n-r)$.

USING TECHNOLOGY EXERCISES, page 377

1. $1.307674368 \times 10^{12}$

3. $2.56094948229 \times 10^{16}$

5. $674,274,182,400$

7. $133,784,560$

9. $4,656,960$

11. Using the multiplication principle, the number of 10-question exams she can set is given by
$$C(25,3) \times C(40,5) \times C(30,2) = 658,337,004,000.$$

CHAPTER 6, REVIEW EXERCISES, page 379

1. $\{3\}$. The set consists of all solutions to the equation $3x - 2 = 7$.

3. $\{4,6,8,10\}$

5. Yes.

7. Yes.

9.

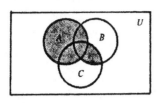

$A \cup (B \cap C)$

11.

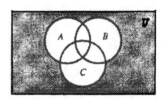

$A^c \cap B^c \cap C^c$

13. $A \cup (B \cup C) = \{a,b\} \cup [\{b,c,d\} \cup \{a,d,e\}]$
$= \{a,b\} \cup \{a,b,c,d,e\} = \{a,b,c,d,e\}.$
$(A \cup B) \cup C = [\{a,b\} \cup \{b,c,d\}] \cup \{a,d,e\}$
$= \{a,b,c,d\} \cup \{a,d,e\} = \{a,b,c,d,e\}.$

15. $A \cap (B \cup C) = \{a,b\} \cap [\{b,c,d\} \cup \{a,d,e\}] = \{a,b\} \cap \{a,b,c,d,e\} = \{a,b\}.$
$(A \cap B) \cup (A \cap C) = [\{a,b\} \cap \{b,c,d\}] \cup [\{a,b\} \cap \{a,d,e\}] = \{b\} \cup \{a\} = \{a,b\}.$

17. The set of all participants in a consumer behavior survey who both avoided buying a product because it is not recyclable and boycotted a company's products because of its record on the environment.

19. The set of all participants in a consumer behavior survey who both did not use cloth diapers rather than disposable diapers and voluntarily recycled their garbage.

In Exercises 21-25, refer to the following Venn diagram.

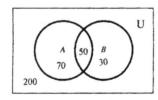

21. $n(A \cup B) = n(A) + n(B) - n(A \cap B) = 120 + 80 - 50 = 150.$

23. $n(B^c) = n(U) - n(B) = 350 - 80 = 270.$

25. $n(A \cap B^c) = n(A) - n(A \cap B) = 120 - 50 = 70.$

27. $C(20,18) = \dfrac{20!}{18!2!} = 190.$

29. $C(5,3) \cdot P(4,2) = \dfrac{5!}{3!2!} \cdot \dfrac{4!}{2!} = 10 \cdot 12 = 120.$

31. Let U denote the set of 5 major cards, and let

   $A = \{ x \in U \mid x \text{ offered cash advances} \}$

   $B = \{ x \in U \mid x \text{ offered extended payments for all goods and services purchased} \}$

   $C = \{ x \in U \mid x \text{ required an annual fee that was less than \$35} \}$

Thus, $n(A) = 3$, $n(B) = 3$, $n(C) = 2$
   $n(A \cap B) = 2$, $n(B \cap C) = 1$
and $n(A \cap B \cap C) = 0$

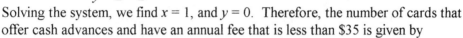

Using the Venn diagram on the right, we have
$$x + y + 2 = 3$$
$$y + 2 = 2$$

Solving the system, we find $x = 1$, and $y = 0$. Therefore, the number of cards that offer cash advances and have an annual fee that is less than \$35 is given by
$$n(A \cap C) = y = 0.$$

33. The number of ways the compact discs can be arranged on a shelf is
$$P(6,6) = 6! = 720 \text{ ways.}$$

35. The number of possible outcomes is $(6)(4)(5)(6) = 720$.

37. a. Since there is repetition of the letters C, I, and N, we use the formula for the permutation of n objects, not all distinct, with $n = 10$, $n_1 = 2$, $n_2 = 3$, and $n_3 = 3$. Then the number of permutations that can be formed is given by
$$\dfrac{10!}{2!3!3!} = 50,400.$$

b. Here, again, we use the formula for the permutation of n objects, not all distinct, this time with $n = 8$, $n_1 = 2$, $n_2 = 2$, and $n_3 = 2$. Then the number of permutations is given by
$$\dfrac{8!}{2!2!2!} = 5040.$$

39. a. The number of ways the 7 students can be assigned to seats is
$$P(7,7) = 7! = 5040.$$
   b. The number of ways 2 specified students can be seated next to each other is
$$2(6) = 12.$$
(Think of seven numbered seats. Then the students can be seated in seats 1-2, or 2-3, or 3-4, or 4-5, or 5-6, or 6-7. Since there are 6 such possibilities and the pair of students can be seated in 2 ways, we conclude that there are 2(6) possible arrangements.)
Then the remaining 5 students can be seated in $P(5,5) = 5!$ ways. Therefore, the number of ways the 7 students can be seated if two specified students sit next to each other is
$$2(6)5! = 1440.$$
Finally, the number of ways the students can be seated if the two students do not sit next to each other is
$$P(7,7) - 2(6)5! = 5040 - 1440 = 3600.$$

41. a. The number of samples that can be selected is
$$C(15,4) = \frac{15!}{4!11!} = 1365.$$
   b. There are
$$C(10,4) = \frac{10!}{4!6!} = 210$$
ways of selecting 4 balls none of which are white. Therefore, there are
$$1365 - 210,$$
or 1155 ways of selecting 4 balls of which at least one is white.

CHAPTER 7

1. $E \cup F = \{a,b,d,f\}$; $E \cap F = \{a\}$.

3. $F^c = \{b,c,e\}$; $E \cap G^c = \{a,b\} \cap \{a,d,f\} = \{a\}$.

5. Since $E \cap F = \{a\}$ is not a null set, we conclude that E and F are not mutually exclusive.

7. $E \cup F \cup G = \{2,4,6\} \cup \{1,3,5\} \cup \{5,6\} = \{1,2,3,4,5,6\}$.

9. $(E \cup F \cup G)^c = \{1,2,3,4,5,6\}^c = \varnothing$.

11. Yes, $E \cap F = \varnothing$; that is, E and F do not contain any common elements.

13. $E \cup F$ 15. G^c

17. $(E \cup F \cup G)^c$

19. $\varnothing$, $\{a\}$, $\{b\}$, $\{c\}$, $\{a,b\}$, $\{a,c\}$, $\{b,c\}$, $\{a,b,c\}$.

21. a. $S = \{R,B\}$ b. $\varnothing$, $\{B\}$, $\{R\}$, $\{B,R\}$

23. a. $S = \{(H,1), (H,2), (H,3), (H,4), (H,5), (H,6), (T,1), (T,2),$
                              $(T,3), (T,4), (T,5), (T,6)\}$
        b. $E = \{(H,2), (H,4), (H,6)\}$

25. $S = \{(d,d,d), (d,d,n), (d,n,d), (n,d,d), (d,n,n), (n,d,n),$
              $(n,n,d), (n,n,n)\}$

27. a. $\{ABC, ABD, ABE, ACD, ACE, ADE, BCD, BCE, BDE, CDE\}$;
        b. 6   c. 3   d. 6

29. a. E^C b. $E^C \cap F^C$ c. $E \cup F$ d. $(E \cap F^C) \cup (E^C \cap F)$

31. a. $S = \{x \mid x > 0\}$ b. $E = \{x \mid 0 < x \le 2\}$
    c. $F = \{x \mid x > 2\}$

33. a. $S = \{0,1,2,3,...,10\}$ b. $E = \{0,1,2,3\}$
    c. $F = \{5,6,7,8,9,10\}$

35. a. $S = \{0,1,2,...,20\}$ b. $E = \{0,1,2,...,9\}$ c. $F = \{20\}$

37. If E is an event of an experiment then E^c is the event containing the elements in S that are not in E. Therefore $E \cap E^C = \varnothing$ and the two sets are mutually exclusive.

39. The number of events of this experiment is 2^n.

41. True. The sample space is $S = \{(1,2), (1,3), (2,1), (2,3), (3,1), (3,2)\}$.

EXERCISES 7.2, page 400

1. $\{(H,H)\}, \{(H,T)\}, \{(T,H)\}, \{(T,T)\}$.

3. $\{(D,m)\}, \{(D,f)\}, \{(R,m)\}, \{(R,f)\}, \{(I,m)\}, \{(I,f)\}$

5. $\{(1,i)\}, \{(1,d)\}, \{(1,s)\}, \{(2,i)\}, \{(2,d)\}, \{(2,s)\}, ..., \{(5,i)\}, \{(5,d)\}, \{(5,s)\}$

7. $\{(A, Rh^+)\}, \{(A, Rh^-)\}, \{(B, Rh^+)\}, \{B, Rh^-)\}, \{(AB, Rh^+)\}, \{(AB, Rh^-)\},$
    $\{(O, Rh^+)\}, \{(O, Rh^-)\}$

9. The probability distribution associated with this data is

Grade	A	B	C	D	F
Probability	0.10	0.25	0.45	0.15	0.05

11. a. $S = \{(0 < x \le 200), (200 < x \le 400), (400 < x \le 600), (600 < x \le 800),$
    $(800 < x \le 1000), (x > 1000)\}$

b.

Number of cars (x)	Probability
$0 < x \le 200$	0.075
$200 < x \le 400$	0.1
$400 < x \le 600$	0.175
$600 < x \le 800$	0.35
$800 < x \le 1000$	0.225
$x > 1000$	0.075

13. The probability distribution associated with this data is

Rating	A	B	C	D	F
Probability	0.026	0.199	0.570	0.193	0.012

15. The probability distribution is

Number of figures produced (in dozens)	30	31	32	33	34	35	36
Probability	0.125	0	0.1875	0.25	0.1875	0.125	0.125

17. The probability is
$$\frac{84{,}000{,}000}{179{,}000{,}000} = 0.469 .$$

19. a. The probability that a person killed by lightning is a male is
$$\frac{376}{439} \approx 0.856 .$$
   b. The probability that a person killed by lightning is a female is
$$\frac{439 - 376}{439} = \frac{63}{439} \approx 0.144 .$$

21. The probability that the retailer uses electronic tags as antitheft devices is
$$\frac{81}{179} \approx 0.46 \ .$$

23. a. $P(D) = \frac{13}{52} = \frac{1}{4}$ b. $P(B) = \frac{26}{52} = \frac{1}{2}$ c. $P(A) = \frac{4}{52} = \frac{1}{13}$

25. The probability of arriving at the traffic light when it is red is
$$\frac{30}{30+5+45} = \frac{30}{80} = 0.375 \ .$$

27. The probability is
$$P(D) + P(C) + P(B) + P(A) = 0.15 + 0.45 + 0.25 + 0.10 = 0.95.$$

29. a. $P(E) = \frac{62}{9+62+27} = \frac{62}{98} \approx 0.633$ b. $P(E) = \frac{27}{98} \approx 0.276$

31. There are two ways of getting a 7, one die showing a 3 and the other die showing a 4, and vice versa.

33. No, the outcomes are not equally likely.

35. No. Since the coin is weighted, the outcomes are not equally likely.

37. a. $P(A) = P(s_1) + P(s_3) = \frac{1}{12} + \frac{1}{12} = \frac{1}{6}$

    b. $P(B) = P(s_2) + P(s_4) + P(s_5) + P(s_6) = \frac{1}{4} + \frac{1}{6} + \frac{1}{3} + \frac{1}{12} = \frac{5}{6}$

    c. $P(C) = 1.$

39. Let G denote a female birth and let B denote a male birth. Then the eight equally likely outcomes of this experiment are
    GGG GGB GBG BGG BGB BBG GBB BBB.
    a. The event that there are two girls and a boy in the family is
       $E = \{GGB, GBG, BGG\}$.
    Since there are three favorable outcomes, $P(E) = 3/8$.
    b. The event that the oldest child is a girl is
       $F = \{GGG, GGB, GBG, GBB\}$.
    Since there are 4 favorable outcomes, $P(F) = 1/2$.

c. The event that the oldest child is a girl and the youngest child is a boy is
$$G = \{GGB, GBB\}.$$

Since there are two favorable outcomes, $P(F) = 1/4$.

41. The probability that the primary cause of the crash was due to pilot error or bad weather is given by $\dfrac{327+22}{327+49+14+22+19+15} = \dfrac{349}{446} \approx 0.782.$

43. True. $P(E) = \dfrac{n(E)}{n(S)} = \dfrac{3}{n}.$

EXERCISES 7.3, page 411

1. Refer to Example 3, page 398. Let E denote the event of interest. Then

$$P(E) = \frac{18}{36} = \frac{1}{2}$$

3. Refer to Example 3, page 398. The event of interest is $E = \{1,1\}$, and $P(E) = 1/36$.

5. Let E denote the event of interest. Then $E = \{(6,2),(6,1),(1,6),(2,6)\}$
   and $\quad P(E) = \dfrac{4}{36} = \dfrac{1}{9}.$

7. Let E denote the event that the card drawn is a king, and let F denote the event that the card drawn is a diamond. Then the required probability is
   $$P(E \cap F) = \frac{1}{52}.$$

9. Let E denote the event that a face card is drawn. Then
   $$P(E) = \frac{12}{52} = \frac{3}{13}.$$

11. Let E denote the event that an ace is drawn. Then $P(E) = 1/13$. Then E^c is the event that an ace is not drawn and
    $$P(E^c) = 1 - P(E) = \frac{12}{13}.$$

13. Let E denote the event that a ticket holder will win first prize, then
$$P(E) = \frac{1}{500} = 0.002,$$
and the probability of the event that a ticket holder will not win first prize is
$$P(E^c) = 1 - 0.002 = 0.998.$$

15. Property 2 of the laws of probability is violated. The sum of the probabilities must add up to 1. In this case $P(S) = 1.1$, which is not possible.

17. The five events are not mutually exclusive; the probability of winning at least one purse is
$$1 - \text{probability of losing all 5 times} = 1 - \frac{9^5}{10^5} = 1 - 0.5905 = 0.4095.$$

19. The two events are not mutually exclusive; hence, the probability of the given event is
$$\frac{1}{6} + \frac{1}{6} - \frac{1}{36} = \frac{11}{36}.$$

21. $E^c \cap F^c = \{c,d,e\} \cap \{a,b,e\} = \{e\} \neq \varnothing$.

23. Let G denote the event that a customer purchases a pair of glasses and let C denote the event that the customer purchases a pair of contact lenses. Then
$$P(G \cup C)^c \neq 1 - P(G) - P(C).$$
Mr. Owens has not considered the case in which the customer buy both glasses and contact lenses.

25. a. $P(E \cap F) = 0$ since E and F are mutually exclusive.
    b. $P(E \cup F) = P(E) + P(F) - P(E \cap F) = 0.2 + 0.5 = 0.7$.
    c. $P(E^c) = 1 - P(E) = 1 - 0.2 = 0.8$.
    d. $P(E^c \cap F^c) = P[(E \cup F)^c] = 1 - P(E \cup F) = 1 - 0.7 = 0.3$.

27. a. $$P(A) = P(s_1) + P(s_2) = \frac{1}{8} + \frac{3}{8} = \frac{1}{2}$$
$$P(B) = P(s_1) + P(s_3) = \frac{1}{8} + \frac{1}{4} = \frac{3}{8}$$
    b. $$P(A^c) = 1 - P(A) = 1 - \frac{1}{2} = \frac{1}{2}$$
$$P(B^c) = 1 - P(B) = 1 - \frac{3}{8} = \frac{5}{8}$$

c. $P(A \cap B) = P(s_1) = \dfrac{1}{8}$

d. $P(A \cup B) = P(A) + P(B) - P(A \cap B) = \dfrac{1}{2} + \dfrac{3}{8} - \dfrac{1}{8} = \dfrac{3}{4}$

29. Referring to the following diagram we see that

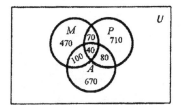

P: lack of parental support
M: malnutrition
A: abused or neglected

the probability that a teacher selected at random from this group said that lack of parental support is the only problem hampering a student's schooling is

$$\dfrac{710}{2140} = 0.33.$$

31. Let E and F denote the events that the person surveyed learned of the products from *Good Housekeeping* and *The Ladies Home Journal*, respectively. Then

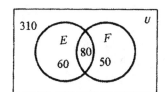

$$P(E) = \dfrac{140}{500} = \dfrac{7}{25}, \quad P(F) = \dfrac{130}{500} = \dfrac{13}{50}$$

and $P(E \cap F) = \dfrac{80}{500} = 0.16$

a. $P(E \cap F) = \dfrac{80}{500} = 0.16$

b. $P(E \cup F) = \dfrac{14}{50} + \dfrac{13}{50} - \dfrac{8}{50} = \dfrac{19}{50} = 0.38$

c. $P(E \cap F^c) + P(E^c \cap F) = \dfrac{60}{500} + \dfrac{50}{500} = \dfrac{110}{500} = 0.22$.

33. Let $A = \{t \,|\, t < 3\}$, $B = \{t \,|\, t \le 4\}$, $C = \{t \,|\, t > 5\}$

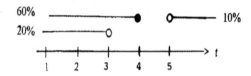

a. $D = \{t \mid t \le 5\}$ and $P(D) = 1 - P(C) = 1 - 0.1 = 0.9$.

b. $E = \{t \mid t > 4\}$ and $P(E) = 1 - P(B) = 1 - 0.6 = 0.4$.

c. $F = \{t \mid 3 \le t \le 4\}$ and $P(F) = P(A^c \cap B) = 0.4$.

35. a. The probability that the participant favors tougher gun-control laws is
$$\frac{150}{250} = 0.6.$$

b. The probability that the participant owns a handgun is
$$\frac{58 + 25}{250} = 0.332.$$

c. The probability that the participant owns a handgun but not a rifle is
$$\frac{58}{250} = 0.232.$$

d. The probability that the participant favors tougher gun-control laws and does not own a handgun is
$$\frac{12 + 138}{250} = 0.6.$$

37. The probability that Bill will fail to solve the problem is $1 - p_1$, and the probability that Mike will fail to solve the problem is $1 - p_2$. Therefore, the probability that both Bill and Mike will fail to solve the problem is $(1 - p_1)(1 - p_2)$. So, the probability that at least one of them will solve the problem is
$$1 - (1 - p_1)(1 - p_2) = 1 - (1 - p_2 - p_1 + p_1 p_2) = p_1 + p_2 - p_1 p_2$$

39. True. Write $B = A \cup (B - A)$. Since A and $B - A$ are mutually exclusive, we have
$$P(B) = P(A) + P(B - A).$$
Since $P(B) = 0$, we have $P(A) + P(B - A) = 0$. If $P(A) > 0$, then $P(B - A) < 0$ and this is not possible. Therefore, $P(A) = 0$.

41. False. Take $E_1 = \{1, 2\}$ and $E_2 = \{2, 3\}$ where $S = \{1, 2, 3\}$. Then

$$P(E_1) = \frac{2}{3}, \ P(E_2) = \frac{2}{3}, \ \text{but} \ P(E_1 \cup E_2) = P(S) = 1.$$

EXERCISES 7.4, page 420

1. Let E denote the event that the coin lands heads all five times. Then

$$P(E) = \frac{1}{2^5} = \frac{1}{32}.$$

3. Let E denote the event that the coin lands tails all 5 times, then

$$P(E^c) = 1 - P(E) = 1 - \frac{1}{32} = \frac{31}{32},$$

    where E^c is the event that the coin lands heads at least once.

5. $P(E) = \dfrac{13 \cdot C(4,2)}{C(52,2)} = \dfrac{78}{1326} \approx 0.0588.$

7. $P(E) = \dfrac{C(26,2)}{C(52,2)} = \dfrac{325}{1326} \approx 0.2451.$

9. The probability of the event that two of the balls will be white and two will be blue is
$$P(E) = \frac{n(E)}{n(S)} = \frac{C(3,2) \cdot C(5,2)}{C(8,4)} = \frac{(3)(10)}{70} = \frac{3}{7}.$$

11. The probability of the event that exactly three of the balls are blue is
$$P(E) = \frac{n(E)}{n(S)} = \frac{C(5,3)C(3,1)}{C(8,4)} = \frac{30}{70} = \frac{3}{7}.$$

13. $$P(E) = \frac{C(3,2) \cdot C(1,1)}{8} = \frac{3}{8}.$$

15. $$P(E) = 1 - \frac{C(3,3)}{8} = \frac{1}{8}.$$

17. The number of elements in the sample space is 2^{10}. There are
$$C(10,6) = \frac{10!}{6!4!},$$

or 210 ways of answering exactly six questions correctly. Therefore, the required probability is

$$\frac{210}{2^{10}} = \frac{210}{1024} \approx 0.205.$$

19. a. Let E denote the event that both of the bulbs are defective. Then

$$P(E) = \frac{C(4,2)}{C(24,2)} = \frac{\dfrac{4!}{2!2!}}{\dfrac{24!}{22!2!}} = \frac{4 \cdot 3}{24 \cdot 23} = \frac{1}{46} \approx 0.022.$$

b. Let F denote the event that none of the bulbs are defective. Then

$$P(F) = \frac{C(20,2)}{C(24,2)} = \frac{20!}{18!2!} \cdot \frac{22!2!}{24!} = \frac{20}{24} \cdot \frac{19}{23} = 0.6884.$$

Therefore, the probability that at least one of the light bulbs is defective is given by
$$1 - P(F) = 1 - 0.6884 = 0.3116.$$

21. a. The probability that both of the cartridges are defective is

$$P(E) = \frac{C(6,2)}{C(80,2)} = \frac{30}{6320} = 0.0048.$$

b. Let F denote the event that none of the cartridges are defective. Then

$$P(F) = \frac{C(74,2)}{C(80,2)} = \frac{5402}{6320} = 0.855,$$

and $P(F^c) = 1 - P(F) = 1 - 0.855 = 0.145$ is the probability that at least 1 of the cartridges is defective.

23. a. The probability that Mary's name will be selected is

$$P(E) = \frac{12}{100} = 0.12;$$

The probability that both Mary's and John's names will be selected is

$$P(F) = \frac{C(98,10)}{C(100,12)} = \frac{\dfrac{98!}{88!10!}}{\dfrac{100!}{88!12!}} = \frac{12 \cdot 11}{100 \cdot 99} \approx 0.013.$$

b. The probability that Mary's name will be selected is

$$P(M) = \frac{6}{40} = 0.15.$$

The probability that both Mary's and John's names will be selected is

$$P(M) \cdot P(J) = \frac{6}{60} \cdot \frac{6}{40} = \frac{36}{2400} = 0.015.$$

25. The probability is given by

$$\frac{C(12,8) \cdot C(8,2)}{C(20,10)} + \frac{C(12,9)C(8,1)}{C(20,10)} + \frac{C(12,10)}{C(20,10)}$$

$$= \frac{(28)(495) + (220)(8) + 66}{184,756} \approx 0.085$$

27. a. The probability that he will select brand B is

$$\frac{C(4,2)}{C(5,3)} = \frac{6}{10} = \frac{3}{5}. \qquad (\frac{\text{the number of selections that include brand } B}{\text{the number of possible selections}})$$

   b. The probability that he will select brands B and C is

$$\frac{C(3,1)}{C(5,3)} = 0.3$$

   c. The probability that he will select at least one of the two brands, B and C is

$$1 - \frac{C(3,3)}{C(5,3)} = 0.9. \qquad \text{(1 - probability that he does not select brands } B \text{ and } C.)$$

29. The probability that the three "Lucky Dollar" symbols will appear in the window of the slot machine is

$$P(E) = \frac{n(E)}{n(S)} = \frac{(1)(1)(1)}{C(9,1)C(9,1)C(9,1)} = \frac{1}{729}.$$

31. The probability of a ticket holder having all four digits in exact order is

$$\frac{1}{C(10,1) \cdot C(10,1) \cdot C(10,1) \cdot C(10,1)} = \frac{1}{10,000} = 0.0001.$$

33. The probability of a ticket holder having a specified digit in exact order is

$$\frac{C(1,1)C(10,1)C(10,1)C(10,1)}{10^4} = 0.10.$$

35. The number of ways of selecting a 5-card hand from 52 cards is given by
$$C(52,5) = 2,598,960.$$
The number of straight flushes that can be dealt in each suit is 10, so there are 4(10)

possible straight flushes. Therefore, the probability of being dealt a straight flush is

$$\frac{4(10)}{C(52,5)} = \frac{40}{2{,}598{,}960} = 0.0000154 \,.$$

37. The number of ways of being dealt a flush in one suit is $C(13,5)$, and, since there are four suits, the number of ways of being dealt a flush is $4 \cdot C(13,5)$. Since we wish to exclude the hands that are straight flushes we subtract the number of possible straight flushes from $4 \cdot C(13,5)$. Therefore, the probability of being drawn a flush, but not a straight flush, is

$$\frac{4 \cdot C(13,5) - 40}{C(52,5)} = \frac{5108}{2{,}598{,}960} = 0.0019654 \,.$$

39. The total number of ways to select three cards of one rank is
$$13 \cdot C(4,3).$$
The remaining two cards must form a pair of another rank and there are
$$12 \cdot C(4,2)$$
ways of selecting these pairs. Next, the total number of ways to be dealt a full house is
$$13 \cdot C(4,3) \cdot 12 \cdot C(4,2) = 3744.$$
Hence, the probability of being dealt a full house is
$$\frac{3{,}744}{2{,}598{,}960} \approx 0.0014406 \,.$$

41. Let E denote the event that in a group of 5, no two will have the same sign. Then
$$P(E) = \frac{12 \cdot 11 \cdot 10 \cdot 9 \cdot 8}{12^5} \approx 0.381944.$$
Therefore, the probability that at least two will have the same sign is given by
$$1 - P(E) = 1 - 0.381944 \approx 0.618.$$
b. $P(\text{no Aries}) = \dfrac{11 \cdot 11 \cdot 11 \cdot 11 \cdot 11}{12^5} \approx 0.647228.$

$$P(1 \text{ Aries}) = \frac{C(5,1) \cdot (1)(11)(11)(11)(11)}{12^5} \approx 0.2941945 \,.$$
Therefore, the probability that at least two will have the sign Aries is given by

$$1 - [P(\text{no Aries}) + P(1 \text{ Aries})] = 1 - 0.9414225 \approx 0.059.$$

43. Referring to the table on page 419, we see that in a group of 50 people, the probability that none of the people will have the same birthday is $1 - 0.970 = 0.03$.

EXERCISES 7.5, page 436

1. a. $P(A|B) = \dfrac{P(A \cap B)}{P(B)} = \dfrac{0.2}{0.5} = \dfrac{2}{5}$.

    b. $P(B|A) = \dfrac{P(A \cap B)}{P(A)} = \dfrac{0.2}{0.6} = \dfrac{1}{3}$.

3. $P(A \cap B) = P(A)P(B|A) = (0.6)(0.5) = 0.3$.

5. $P(A) \cdot P(B) = (0.3)(0.6) = 0.18 = P(A \cap B)$. Therefore the events are independent.

7. $P(A \cap B) = P(A) + P(B) - P(A \cup B) = 0.5 + 0.7 - 0.85 = 0.35$
    $$= P(A) \cdot P(B)$$
    so they are independent events.

9. a. $P(A \cap B) = P(A)P(B) = (0.4)(0.6) = 0.24$.
    b. $P(A \cup B) = P(A) + P(B) - P(A \cap B) = 0.4 + 0.6 - 0.24 = 0.76$.

11. a. $P(A) = 0.5$ b. $P(E|A) = 0.4$
    c. $P(A \cap E) = P(A)P(E|A) = (0.5)(0.4) = 0.2$
    d. $P(E) = (0.5)(0.4) + (0.5)(0.3) = 0.35$.
    e. No. $P(A \cap E) \neq P(A) \cdot P(E) = (0.5)(0.35)$
    f. A and E are not independent events.

13.

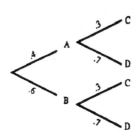

    a. $P(A) = 0.4$ b. $P(C|A) = 0.3$
    c. $P(A \cap C) = P(A)P(C|A) = (0.4)(0.3) = 0.12$

d. $P(C) = (0.4)(0.3) + (0.6)(0.3) = 0.3$
e. Yes. $P(A \cap C) = 0.12 = P(A)P(C)$
f. Yes.

15. Let A denote the event that the sum of the numbers is less than 9 and let B denote the event that one of the numbers is a 6.

Then, $P(A|B) = \dfrac{P(A \cap B)}{P(B)} = \dfrac{\frac{4}{36}}{\frac{11}{36}} = \dfrac{4}{11}$.

17. Refer to Figure 7.15, page 426 in the text. Here
$$E = \{(3,1), (3,2), (3,3), (3,4), (3,5), (3,6)\}$$
and $F = \{(1,6), (6,1), (2,5), (5,2), (3,4), (4,3)\}$.
Then $E \cap F = \{(3,4)\}$.

Now, $P(E \cap F) = \dfrac{1}{36}$ and this is equal to $P(E) \cdot P(F) = \left(\dfrac{6}{36}\right)\left(\dfrac{6}{36}\right) = \dfrac{1}{36}$.

So E and F are independent events.

19. Let A denote the event that the battery lasts 10 or more hours and let B denote the event that the battery lasts 15 or more hours.
Then $P(A) = 0.8, \quad P(B) = 0.15$
and $P(A \cap B) = 0.15$.

Therefore, the probability that the battery will last 15 hours or more is
$$P(B|A) = \dfrac{P(A \cap B)}{P(A)} = \dfrac{0.15}{0.8} = \dfrac{3}{16} = 0.1875.$$

21. Refer to the following tree diagram:

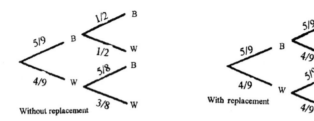

a. The probability that the second ball drawn is a white ball if the second ball is drawn without replacing the first is

$$P(B)P(W|B) + P(W)P(W|W) = (\frac{5}{9})(\frac{1}{2}) + (\frac{4}{9})(\frac{3}{8}) = \frac{4}{9}.$$

b. The probability that the second ball drawn is a white ball if the first ball is replaced before the second is drawn is

$$(\frac{5}{9})(\frac{4}{9}) + (\frac{4}{9})(\frac{4}{9}) = \frac{4}{9}.$$

23. Refer to the following tree diagram:

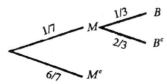

a. The probability that a student selected at random from this medical school is black is

$$(\frac{1}{7})(\frac{1}{3}) = \frac{1}{21}$$

b. The probability that a student selected at random from this medical school is black if it is known that the student is a member of a minority group is $P(B|M) = 1/3$.

25. Let D denote the event that the card drawn is a diamond. Consider the tree diagram that follows.

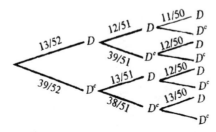

Then the required probability is

$$(\frac{13}{52})(\frac{12}{51})(\frac{11}{50}) + (\frac{13}{52})(\frac{39}{51})(\frac{12}{50}) + (\frac{39}{52})(\frac{13}{51})(\frac{12}{50}) + (\frac{39}{52})(\frac{38}{51})(\frac{13}{50})$$
$$= 0.25.$$

27. The sample space for a three-child family is
$$S = \{GGG, GGB, GBG, GBB, BGG, BGB, BBG, BBB\}.$$
Since we know that there is at least one girl in the three-child family we are dealing

with a reduced sample space
$$S_1 = \{GGG, GGB, GBG, GBB, BGG, BGB, BBG\}$$
in which there are 7 outcomes. Then the probability that all three children are girls is
$$P(E) = \frac{n(E)}{n(S)} = \frac{1}{7}.$$

29. a.

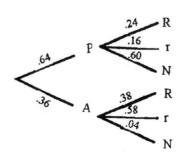

P = Professional
A = Amateur
R = recovered within 48 hrs.
r = recovered after 48 hrs.
N = never recovered

b. The required probability is 0.24.
c. The required probability is $(0.64)(0.60) + (0.36)(0.04) \approx 0.40$.

31. Refer to the following tree diagram.

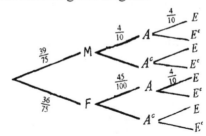

a. $P(A \cap E | M) = (0.4)(0.4) = 0.16.$

b. $P(A) = P(M \cap A) + P(F \cap A) = \frac{39}{75} \cdot \frac{4}{10} + \frac{36}{75} \cdot \frac{45}{100} = 0.424$

c. $P(M \cap A \cap E) + P(F \cap A \cap E) = \frac{39}{75} \cdot \frac{4}{10} \cdot \frac{4}{10} + \frac{36}{75} \cdot \frac{45}{100} \cdot \frac{4}{10}$

$$= 0.0832 + 0.0864 = 0.1696.$$

33. a. The probability that none of the dozen eggs is broken is
$$(0.992)^{12} = 0.908.$$

Therefore, the probability that at least one egg is broken is
$$1 - 0.908 = 0.092.$$
b. Using the results of (a), we see that the required probability is
$$(0.092)(0.092)(0.908) \approx 0.008.$$

35. a. $P(A) = \dfrac{1120}{4000} = 0.280$; $P(B) = \dfrac{1560}{4000} = 0.390$;

$$P(A \cap B) = \dfrac{720}{4000} = 0.180$$

$$P(B|A) = \dfrac{P(A \cap B)}{P(A)} = \dfrac{n(A \cap B)}{n(A)} = \dfrac{720}{1120} \approx 0.643$$

$$P(B|A^C) = \dfrac{P(A^C \cap B)}{P(A^C)} = \dfrac{n(A^C \cap B)}{n(A^C)} = \dfrac{840}{2880} \approx 0.292.$$

b. $P(B|A) \neq P(B)$ so A and B are not independent events.

37. Let C denote the event that a person in the survey was a heavy coffee drinker and Pa denote the event that a person in the survey had cancer of the pancreas. Then

$$P(C) = \dfrac{3200}{10000} = 0.32 \text{ and } P(Pa) = \dfrac{160}{10000} = 0.016$$

$$P(C \cap Pa) = \dfrac{132}{160} = 0.825$$

and $P(C) \cdot P(Pa) = 0.00512 \neq P(C \cap Pa)$. Therefore the events are not independent.

39. The probability that the first test will fail is 0.03, that the second test will fail is 0.015, and the third test will fail is 0.015. Since these are independent events the probability that all three tests will fail is
$$(0.03)(0.015)(0.015) = 0.0000068.$$

41. a. Let $P(A)$, $P(B)$, $P(C)$ denote the probability that the first, second, and third patient suffer a rejection, respectively. Then $P(A) = \dfrac{1}{2}$, $P(B) = \dfrac{1}{3}$, and $P(C) = \dfrac{1}{10}$.

Therefore, the probabilities that each patient does not suffer a rejection are given by $P(A^c) = \dfrac{1}{2}$, $P(B^c) = \dfrac{2}{3}$, and $P(C^c) = \dfrac{9}{10}$. Then, the probability that none of the 3 patients suffers a rejection is given by

$$P(A^c) \cdot P(B^c) \cdot P(C^c) = \dfrac{1}{2} \cdot \dfrac{2}{3} \cdot \dfrac{9}{10} = \dfrac{18}{60} = \dfrac{3}{10}.$$

Therefore, the probability that at least one patient will suffer rejection is

$$1 - P(A^c) \cdot P(B^c) \cdot P(C^c) = 1 - \frac{3}{10} = \frac{7}{10}.$$

b. The probability that exactly two patients will suffer rejection is
$$P(A)P(B)P(C^c) + P(A)P(B^c)P(C) + P(A^c)P(B)P(C)$$
$$= \frac{1}{2} \cdot \frac{1}{3} \cdot \frac{9}{10} + \frac{1}{2} \cdot \frac{2}{3} \cdot \frac{1}{10} + \frac{1}{2} \cdot \frac{1}{3} \cdot \frac{1}{10} = \frac{9+2+1}{60} = \frac{12}{60} = \frac{1}{5}.$$

43. Let A denote the event that at least one of the floodlights remain functional over the one-year period. Then
$$P(A) = 0.99999 \text{ and } P(A^c) = 1 - P(A) = 0.00001.$$
Letting n represent the minimum number of floodlights needed, we have
$$(0.01)^n = 0.00001$$
$$n\log(0.01) = -5$$
$$n(-2) = -5$$
$$n = \tfrac{5}{2} = 2.5.$$
Therefore, the minimum number of floodlights needed is 3.

45. The probability that the event will not occur in one trial is $1 - p$. Therefore, the probability that it will not occur in n independent trials is $(1 - p)^n$. Therefore, the probability that it will occur at least once in n independent trials is $1 - (1 - p)^n$.

47. True. Since A and B are mutually exclusive, $A \cap B = \varnothing$, and
$$P(A|B) = \frac{P(A \cap B)}{P(B)} = \frac{P(\varnothing)}{P(B)} = 0.$$

49. True. Since $A \cap B = \varnothing$, $P(A \cap B) = 0$. Therefore,
$$P(A|A \cup B) = \frac{P[A \cap (A \cup B)]}{P(A \cup B)} = \frac{P(A)}{P(A) + P(B) + P(A \cap B)} = \frac{P(A)}{P(A) + P(B)}.$$

EXERCISES 7.6, page 446

1.

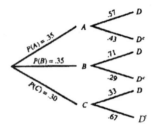

3. a. $P(D^C) = \dfrac{15+10+20}{35+35+30} = 0.45$ b. $P(B|D^C) = \dfrac{10}{15+10+20} = 0.22$.

5. a. $P(D) = \dfrac{25+20+15}{50+40+35} = 0.48$ b. $P(B|D) = \dfrac{20}{25+20+15} = 0.33$

7. a. $P(A) \cdot P(D|A) = (0.4)(0.2) = 0.08$
    b. $P(B) \cdot P(D|B) = (0.6)(0.25) = 0.15$
    c. $P(A|D) = \dfrac{P(A) \cdot P(D|A)}{P(A) \cdot P(D|A) + P(B) \cdot P(D|B)} = \dfrac{(0.4)(0.2)}{0.08+0.15} \approx 0.35$

9. a. $P(A) \cdot P(D|A) = \dfrac{1}{3} \cdot \dfrac{1}{4} = \dfrac{1}{12}$ b. $P(B) \cdot P(D|B) = \dfrac{1}{2} \cdot \dfrac{1}{2} = \dfrac{1}{4}$

    c. $P(C) \cdot P(D|C) = \dfrac{1}{6} \cdot \dfrac{1}{3} = \dfrac{1}{18}$

    d. $P(A|D) = \dfrac{P(A) \cdot P(D|A)}{P(A) \cdot P(D|A) + P(B) \cdot P(D|B) + P(C) \cdot P(C|B)}$

    $= \dfrac{\frac{1}{12}}{\frac{1}{12} + \frac{1}{4} + \frac{1}{18}} = \dfrac{1}{12} \cdot \dfrac{36}{14} = \dfrac{3}{14}$

11. Let A denote the event that the first card drawn is a heart and B the event that the second card drawn is a heart. Then

    $$P(A|B) = \dfrac{P(A) \cdot P(B|A)}{P(A) \cdot P(B|A) + P(A^C) \cdot P(B|A^C)}$$

    $$= \dfrac{\frac{1}{4} \cdot \frac{12}{51}}{\frac{1}{4} \cdot \frac{12}{51} + \frac{3}{4} \cdot \frac{13}{51}} = \dfrac{4}{17}.$$

13. Using the tree diagram at the right, we see that

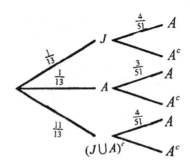

$$P(J|A) = \frac{\frac{1}{13} \cdot \frac{4}{51}}{\frac{1}{13} \cdot \frac{4}{51} + \frac{1}{13} \cdot \frac{3}{51} + \frac{11}{13} \cdot \frac{4}{51}} = \frac{\frac{4}{13 \cdot 51}}{\frac{51}{13 \cdot 51}} = 0.0784 \ .$$

15. The probabilities associated with this experiment are represented in the following tree diagram.

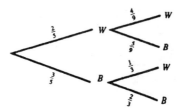

17. Referring to the tree diagram in Exercise 15, we see that the probability that the transferred ball was black given that the second ball was white is

$$P(B|W) = \frac{\frac{3}{5} \cdot \frac{1}{3}}{\frac{2}{5} \cdot \frac{4}{9} + \frac{3}{5} \cdot \frac{1}{3}} = \frac{9}{17} \ .$$

19. Let D denote the event that a senator selected at random is a Democrat, R denote the event that a senator selected at random is a Republican, and M the event that a senator has served in the military. From the following tree diagram

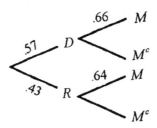

we see that the probability that a senator selected at random who has served in the military is a Republican is

$$P(R|M) = \frac{P(R)P(M|R)}{P(M)} = \frac{(0.64)(0.43)}{(0.66)(0.57) + (0.64)(0.43)}$$

$$\approx 0.422.$$

21. Let H_2 denote the event that the coin tossed is the two-headed coin, H_B denote the event that the coin tossed is the biased coin, and H_F denote the event that the coin tossed is the fair coin. Referring to the following tree diagram, we see that

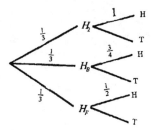

a. $P(H) = \dfrac{1}{3} \cdot 1 + \dfrac{1}{3} \cdot \dfrac{3}{4} + \dfrac{1}{3} \cdot \dfrac{1}{2} = \dfrac{1}{3} + \dfrac{1}{4} + \dfrac{1}{6} = \dfrac{9}{12} = \dfrac{3}{4}$

b. $P(H_F \mid H) = \dfrac{\frac{1}{3} \cdot \frac{1}{2}}{\frac{3}{4}} = \dfrac{2}{9}$

23. Let D denote the event that the person has the disease, and let Y denote the event that the test is positive. Referring to the following tree diagram, we see that the required probability is

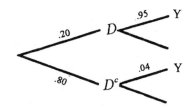

$$P(D \mid Y) = \frac{P(D) \cdot P(Y \mid D)}{P(D) \cdot P(Y \mid D) + P(D^C) \cdot P(Y \mid D^C)}$$

$$= \frac{(0.2)(0.95)}{(0.2)(0.95) + (0.8)(0.04)} \approx 0.856.$$

25. Let x denote the age of an insured driver, and let A denote the event that an insured driver is in an accident. Using the tree diagram we find,

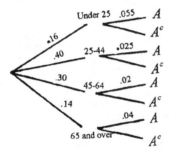

a. $P(A) = (0.16)(0.055) + (0.4)(0.025) + (0.3)(0.02) + (0.14)(0.04)$
≈ 0.03.

b. $P(x < 25 | A) = \dfrac{(0.16)(0.055)}{(0.03)} \approx 0.29$

27. Let E, F, and G denote the events that the child selected at random is 12 years old, 13 years old, or 14 years old, respectively; and let C^c denote the event that the child does not have a cavity. Using the tree diagram at the right, we see that

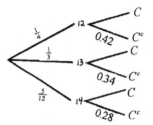

$$P(G|C^c) = \dfrac{\frac{5}{12}(0.28)}{\frac{1}{4}(0.42) + \frac{1}{3}(0.34) + \frac{5}{12}(0.28)} = 0.348$$

29. Let M and F denote the events that a person arrested for crime in 1988 was male or female, respectively; and let U denote the event that the person was under the age of 18. Using the following tree diagram, we have

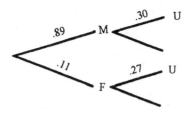

a. $P(U) = (0.89)(0.30) + (0.11)(0.27) = 0.2967$.

b. $P(F|U) = \dfrac{(0.11)(0.27)}{(0.89)(0.30)+(0.11)(0.27)} = 0.1001$.

31. Let I and II denote a customer who purchased the drug in capsule or tablet form, respectively; and let E denote a customer in Group I and II who purchased the extra-strength dosage of the drug. Then using the tree diagram that follows, we see that

$$P(I|E) = \dfrac{(0.57)(0.38)}{(0.57)(0.38)+(0.43)(0.31)} = 0.619$$

33. Let D and N denote the events that a employee was placed by Ms. Dwyer or Ms. Newberg, respectively; and let S denote the event that the employee placed by one of these women was satisfactory. Using the tree diagram shown at the right, we see that

$$P(N|S^C) = \dfrac{(0.55)(0.3)}{(0.45)(0.2)+(0.55)(0.3)}$$

$$= 0.647$$

35. Using the tree diagram shown at the right, we see that

$$P(S_2|S) = \dfrac{(0.95)(0.8)}{(0.95)(0.8)+(0.2)(0.3)} \approx 0.93.$$

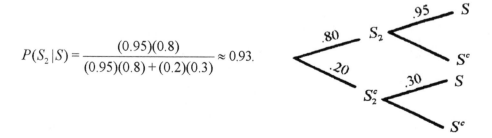

37. a. Let A, B, C, D, and E denote the events that the annual household income is less than 15,000, between 15,000 and 29,999, ..., 75,000, and higher, respectively. Let R, M, and P denote the probability that a person considers himself or herself rich, middle class, or poor, respectively. From the following tree diagram,

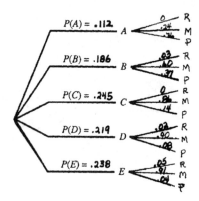

we see that the probability that a respondent chosen at random calls himself or herself middle class is

$$(.112)(.24) + (.186)(.60) + (.245)(.86) + (.219)(.90) + (.238)(.91) = .76286$$

b. $P(C|M) = \dfrac{(.245)(.86)}{(.112)(.24) + (.186)(.6) + (.245)(.86) + (.219)(.9) + (.238)(.91)}$

$\approx .276.$

c. Using the results of (b), the required probability is 1 - 0.276, or 0.724.

CHAPTER 7, REVIEW EXERCISES, page 452

1. a. $P(E \cap F) = 0$ since E and F are mutually exclusive.

b. $P(E \cup F) = P(E) + P(F) - P(E \cap F)$
   $= 0.4 + 0.2 = 0.6$

c. $P(E^c) = 1 - P(E) = 1 - 0.4 = 0.6.$

d. $P(E^c \cap F^c) = P(E \cup F)^c = 1 - P(E \cup F) = 1 - 0.6 = 0.4.$

e. $P(E^c \cup F^c) = P(E \cap F)^c = 1 - P(E \cap F) = 1 - 0 = 1.$

3. a. The probability of the number being even is
$$P(2) + P(4) + P(6) = 0.12 + 0.18 + 0.19 = 0.49.$$

b. The probability that the number is either a 1 or a 6 is
$$P(1) + P(6) = 0.20 + 0.19 = 0.39.$$
c. The probability that the number is less than 4 is
$$P(1) + P(2) + P(3) = 0.20 + 0.12 + 0.16 = 0.48.$$

5. Let A denote the event that a video-game cartridge has an audio defect and let V denote the event that a video-game cartridge has a video defect. Then using the following Venn diagram, we have

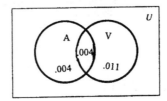

a. the probability that a cartridge purchased by a customer will have a video or audio defect is
$$P(A \cup V) = P(A \cap V^{c}) + P(V \cap A^{c}) + P(A \cap V)$$
$$= 0.004 + 0.004 + 0.011 = 0.019$$
b. the probability that a cartridge purchased by a customer will have not have a video or audio defect is
$$P(A \cup V)^{c} = 1 - P(A \cup V) = 1 - 0.019 = 0.981.$$

7. $P(A \cap E) = (0.3)(0.6) = 0.18.$ 9. $P(C \cap E) = (0.2)(0.3) = 0.06.$

11. $P(E) = 0.18 + 0.25 + 0.06 = 0.49$

13. a. The probability that none of the pens in the sample are defective is
$$\frac{C(18,3)}{C(20,3)} = \frac{\dfrac{18!}{15!3!}}{\dfrac{20!}{17!3!}} = \frac{18!}{15!} \cdot \frac{17!}{20!} = \frac{68}{95} \approx 0.71579.$$
Therefore, the probability that at least one is defective is given by
$$1 - 0.71579 \approx 0.284.$$

b. The probability that two are defective is given by

$$\frac{C(2,2) \cdot C(18,1)}{C(20,3)} = \frac{18}{\dfrac{20!}{17!3!}} = \frac{6}{380} \approx 0.0158 .$$

Therefore, the probability that no more than 1 is defective is given by

$$1 - \frac{6}{380} = \frac{374}{380} \approx 0.984 .$$

15. Let E denote the event that the sum of the numbers is 8 and let D denote the event that the numbers appearing on the face of the two dice are different. Then

$$P(E|D) = \frac{P(E \cap D)}{P(D)} = \frac{4}{30} = \frac{2}{15}.$$

17. The probability that all three cards are face cards is

$$\frac{C(12,3)}{C(52,3)} = 0.00995.$$

19. Referring to the following tree diagram,

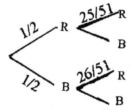

we see that the probability that the second card is black, given that the first card was red is

$$\frac{26}{51} = 0.510 .$$

21. Let M denote the event that an employee at the insurance company is a male and let F denote the event that an employee at the insurance company is on flex time. Then

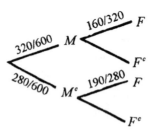

$$P(M|F) = \frac{\dfrac{320}{600} \cdot \dfrac{160}{320}}{\dfrac{320}{600} \cdot \dfrac{160}{320} + \dfrac{280}{600} \cdot \dfrac{190}{280}} = 0.4571.$$

23. Let E denote the event that an applicant selected at random is eligible for admission, and let Pa denote the event that an applicant selected at random passes the admission exam. Using the tree diagram that follows, we see that

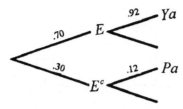

a. $P(Pa) = (0.70)(0.92) + (0.30)(0.12) = 0.68$.

b. $P(E^c | Pa) = \dfrac{(0.30)(0.12)}{(0.70)(0.92) + (0.30)(0.12)} = 0.053$.

CHAPTER 8

EXERCISES 8.1, page 463

1. a. See part (b).
    b.   c. $\{GGG\}$

Outcome	GGG	GGR	GRG	RGG
Value	3	2	2	2

Outcome	GRR	RGR	RRG	RRR
Value	1	1	1	0

3. X may assume the values in the set $S = \{1,2,3,...\}$.

5. The event that the sum of the dice is 7 is
    $$E = \{(1,6),(2,5),(3,4),(4,3),(5,2),(6,1)\}$$

    and $P(E) = \dfrac{6}{36} = \dfrac{1}{6}$.

7. X may assume the value of any positive integer. The random variable is infinite

    discrete.

9. $\{d \mid d \geq 0\}$. The random variable is continuous.

11. X may assume the value of any positive integer. The random variable is infinite
    discrete.

13. a. $P(X = -10) = 0.20$
    b. $P(X \geq 5) = 0.1 + 0.25 + 0.1 + 0.15 = 0.60$
    c. $P(-5 \leq X \leq 5) = 0.15 + 0.05 + 0.1 = 0.30$

d. $P(X \leq 20) = 0.20 + 0.15 + 0.05 + 0.1 + 0.25 + 0.1 + 0.15 = 1$

15.

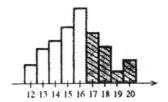

12 13 14 15 16 17 18 19 20

17.   a.

x	1	2	3	4	5	6
$P(X = x)$	$\frac{1}{6}$	$\frac{1}{6}$	$\frac{1}{6}$	$\frac{1}{6}$	$\frac{1}{6}$	$\frac{1}{6}$

y	1	2	3	4	5	6
$P(Y = y)$	$\frac{1}{6}$	$\frac{1}{6}$	$\frac{1}{6}$	$\frac{1}{6}$	$\frac{1}{6}$	$\frac{1}{6}$

b.

$x + y$	2	3	4	5	6	7	8	9	10	11	12
$P(X + Y) = x + y$	$\frac{1}{36}$	$\frac{2}{36}$	$\frac{3}{36}$	$\frac{4}{36}$	$\frac{5}{36}$	$\frac{6}{36}$	$\frac{5}{36}$	$\frac{4}{36}$	$\frac{3}{36}$	$\frac{2}{36}$	$\frac{1}{36}$

19.   a.

x	0	1	2	3	4	5
$P(X=x)$	0.017	0.067	0.033	0.117	0.233	0.133

x	6	7	8	9	10
$P(X=x)$	0.167	0.1	0.05	0.067	0.017

b.

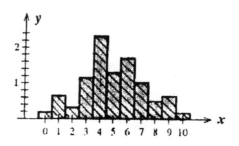

21.

x	1	2	3	4	5
P(X=x)	0.007	0.029	0.021	0.079	0.164

x	6	7	8	9	10
P(X=x)	0.15	0.20	0.207	0.114	0.029

23.   False. The area is exactly equal to one.

USING TECHNOLOGY EXERCISES, page 465

1.   Histogram representing data in Table 3, page 459 in text.

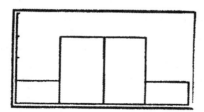

3.

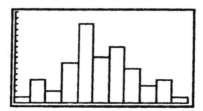

1. a. The student's grade-point average is given by
$$\frac{(2)(4)(3)+(3)(3)(3)+(4)(2)(3)+(1)(1)(3)}{(10)(3)}$$
or 2.6.

    b.

x	0	1	2	3	4
$P(X = x)$	0	0.1	0.4	0.3	0.2

$E(X) = 1(0.1) + 2(0.4) + 3(0.3) + 4(0.2) = 2.6$.

3. $E(X) = -5(0.12) + -1(0.16) + 0(0.28) + 1(0.22) + 5(0.12) + 8(0.1)$
    $= 0.86$.

5. $E(X) = 0(0.07) + 25(0.12) + 50(0.17) + 75(0.14) + 100(0.28)$
    $+ 125(0.18) + 150(0.04)$
    $= 78.5$, or $78.50.

7. A customer entering the store is expected to buy
    $E(X) = (0)(0.42) + (1)(0.36) + (2)(0.14) + (3)(0.05) + (4)(0.03)$
    $= 0.91$, or 0.91 videocassettes.

9. The expected number of accidents is given by
    $E(X) = (0)(0.935) + (1)(0.03) + (2)(0.02) + (3)(0.01) + (4)(0.005)$
    $= 0.12$.

11. The expected number of machines that will break down on a given day is given by
    $E(X) = (0)(0.43) + (1)(0.19) + (2)(0.12) + (3)(0.09) + (4)(0.04)$
    $+ (5)(0.03) + (6)(0.03) + (7)(0.02) + (8)(0.05)$
    $= 1.73$.

13. The expected gain of the insurance company is given by
    $$E(X) = 0.96P - (20,000 - P)(0.04),$$
    where P is the amount the man can expect to pay. We want $E(X) = 0$, that is,
    $$P - (0.04)(20,000) = 0.$$
    Solving for P, we find $P = 800$ and so the minimum he can expect to pay is $800.

15. The expected gain of the insurance company is given by
$$E(X) = 0.9935P - (25{,}000 - P)0.0065 \geq 0,$$
where P is the amount that Dennis can expect to pay. Since $E(X) \geq 0$,
$$P - 0.0065(25{,}000) \geq 0.$$
Solving for P, we find $P \geq 162.50$, so the minimum Dennis can expect to pay is $162.50.

17. The expected value of the first project is
$(0.7)(180{,}000) + (0.3)(150{,}000) = 171{,}000$, or $171,000
and the expected value of the second project is
$(0.6)(220{,}000) + (0.4)(80{,}000) = 164{,}000$, or $164,000.
The proprietor should choose the first project if he wants to maximize his expected profit.

19. a. DAHL MOTORS
$$E(X) = 5(0.05) + 6(0.09) + 7(0.14) + 8(0.24) + 9(0.18)$$
$$+ 10(0.14) + 11(0.11) + 12(0.05)$$
$$= 8.52.$$

FARTHINGTON AUTO SALES
$$E(X) = 5(0.08) + 6(0.21) + 7(0.31) + 8(0.24) + 9(0.10)$$
$$+ 10(0.06)$$
$$= 7.25.$$
b. The expected weekly profit from Dahl Motors is
$$8.52(362) = 3084.24,$$
or $3084.24. The expected weekly profit from Farthington Auto Sales is
$$7.25(436), \quad \text{or } \$3161.$$
We conclude that Roger should purchase Farthington Auto Sales.

21. a. The odds that it will rain tomorrow are 3 to 7.
b. The odds that it will not rain tomorrow are 7 to 3.

23. The expected value of the player's winnings are given by
$$E(X) = (1)(\frac{18}{38}) + (-1)(\frac{18}{38}) + (-2)(\frac{2}{38})$$
$$= -\frac{4}{38} \approx -0.105,$$
or a loss of $10\frac{1}{2}$ cents.

25. The odds in favor of E occurring are
$$\frac{P(E)}{P(E^C)} = \frac{0.8}{0.2},$$
or 4 to 1. The odds against E occurring are 1 to 4.

27. The probability of E occurring is given by
$$P(E) = \frac{9}{9+7} = \frac{9}{16} \approx 0.5625.$$

29. The probability that Carmen will make the sale is given by
$$P(E) = \frac{8}{8+5} = \frac{8}{13} \approx 0.6154.$$

31. The probability that the boxer will win the match is given by
$$P(E) = \frac{4}{3+4} = \frac{4}{7} \approx 0.5714.$$

33. a. $E(cx) = (cx_1)p_1 + (cx_2)p_2 + \cdots + (cx_n)p_n$
$$= c(x_1p_1 + x_2p_2 + \cdots + x_np_n)$$
$$= cE(X).$$
b. The expected loss is $(300)(-\frac{2}{38}) = -15.79$, or a loss of \$15.79.

35. The mean-wage rate is given by

$$(\tfrac{60}{450})(10.70) + (\tfrac{90}{450})(10.80) + (\tfrac{75}{450})(10.90) + (\tfrac{120}{450})(11.00) + (\tfrac{60}{450})(11.10) + (\tfrac{45}{450})(11.20)$$
$$= 10.94$$
or \$10.94. The mode is \$11.00, and the median is

$$\frac{10.90 + 11.00}{2} = 10.95, \text{ or } \$10.95.$$

37. True. This follows from the defintion.

EXERCISES 8.3, page 490

1. $\mu = (1)(0.4) + (2)(0.3) + 3(0.2) + (4)(0.1) = 2.$
Var $(X) = (0.4)(1 - 2)^2 + (0.3)(2 - 2)^2 + (0.2)(3 - 2)^2 + (0.1)(4 - 2)^2$
$$= 0.4 + 0 + 0.2 + 0.4 = 1$$
$\sigma = \sqrt{1} = 1.$

3. $\mu = -2(\frac{1}{16}) + -1(\frac{4}{16}) + 0(\frac{6}{16}) + 1(\frac{4}{16}) + 2(\frac{1}{16}) = \frac{0}{16} = 0$.

Var $(X) = \frac{1}{16}(-2-0)^2 + \frac{4}{16}(-1-0)^2 + \frac{6}{16}(0-0)^2 + \frac{4}{16}(1-0)^2 + \frac{1}{16}(2-0)^2$

$\quad = 1$

$\sigma = \sqrt{1} = 1$.

5. $\mu = 0.1(430) + (0.2)(480) + (0.4)(520) + (0.2)(565) + (0.1)(580)$

$\quad = 518$.

Var $(X) = 0.1(430 - 518)^2 + (0.2)(480 - 518)^2 + (0.4)(520 - 518)^2$

$\qquad + (0.2)(565 - 518)^2 + (0.1)(580 - 518)^2$

$\quad = 1891$.

$\sigma = \sqrt{1891} \approx 43.49$.

7. The mean of the histogram in Figure (b) is more concentrated about its mean than the histogram in Figure (a). Therefore, the histogram in Figure (a) has the larger variance.

9. $E(X) = 1(0.1) + 2(0.2) + 3(0.3) + 4(0.2) + 5(0.2) = 3.2$.

Var $(X) = (0.1)(1 - 3.2)^2 + (0.2)(2 - 3.2)^2 + (0.3)(3 - 3.2)^2$

$\qquad + (0.2)(4 - 3.2)^2 + (0.2)(5 - 3.2)^2$

$\quad = 1.56$

11. $\mu = \dfrac{1+2+3+ \cdots +8}{8} = 4.5$

$V(X) = \frac{1}{8}(1-4.5)^2 + \frac{1}{8}(2-4.5)^2 + \cdots + \frac{1}{8}(8-4.5)^2 = 5.25$

13. a. Let X be the annual birth rate during the years 1981 - 1990.

b.

x	15.5	15.6	15.7	15.9	16.2	16.7
$P(X=x)$	0.2	0.1	0.3	0.2	0.1	0.1

c. $E(X) = (0.2)(15.5) + (0.1)(15.6) + (0.3)(15.7) + (0.2)(15.9)$

$\qquad + (0.1)(16.2) + (0.1)(16.7)$

$\quad = 15.84$.

$V(X) = (0.2)(15.5 - 15.84)^2 + (0.1)(15.6 - 15.84)^2$

$\qquad + (0.3)(15.7 - 15.84)^2 + (0.2)(15.9 - 15.84)^2$

$\qquad + (0.1)(16.2 - 15.84)^2 + (0.1)(16.7 - 15.84)^2$

$$= 0.1224$$

$$\sigma = \sqrt{0.1224} \approx 0.350.$$

15. a. Mutual Fund *A*

$$\mu = (0.2)(-4) + (0.5)(8) + (0.3)(10) = 6.2, \text{ or } \$620.$$
$$V(X) = (0.2)(-4 - 6.2)^2 + (0.5)(8 - 6.2)^2 + (0.3)(10 - 6.2)^2$$
$$= 26.76, \text{ or } \$267,600.$$

Mutual Fund *B*

$$\mu = (0.2)(-2) + (0.4)(6) + (0.4)(8)$$
$$= 5.2, \text{ or } \$520.$$
$$V(X) = (0.2)(-2 - 5.2)^2 + (0.4)(6 - 5.2)^2 + (0.4)(8 - 5.2)^2$$
$$= 13.76, \text{ or } \$137,600.$$

b. Mutual Fund *A*　　　　　　　　c. Mutual Fund *B*

17. $\text{Var}(X) = (0.4)(1)^2 + (0.3)(2)^2 + (0.2)(3)^2 + (0.1)(4)^2 - (2)^2 = 1.$

19. $\mu = [\frac{10}{500}(180) + \frac{20}{500}(190) + \cdots + \frac{5}{500}(350)] = 239.6, \text{ or } \$239,600.$

$$V(X) = [\frac{10}{500}(180 - 239.6)^2 + \frac{20}{500}(190 - 239.6)^2 + \cdots + \frac{5}{500}(350 - 239.6)^2][(1000)^2]$$

$$= 1443.84 \times 10^6 \text{ dollars.}$$

$$\sigma = \sqrt{1443.84 \times 10^6} = 37.998 \times 10^3, \text{ or } \$37,998.$$

21. a. Using Chebychev's inequality we have

$$P(\mu - k\sigma \leq X \leq \mu + k\sigma) \geq 1 - 1/k^2.$$
$$\mu - k\sigma = 42 - k(2) = 38, \text{ and } k = 2,$$

and $\quad P(\mu - k\sigma \leq X \leq \mu + k\sigma) \geq 1 - 1/(2)^2$
$$\geq 1 - 1/4$$
$$\geq 3/4, \text{ or at least } 0.75.$$

b. Using Chebychev's inequality we have

$$P(\mu - k\sigma \leq X \leq \mu + k\sigma) \geq 1 - 1/k^2.$$
$$\mu - k\sigma = 42 - k(2) = 32, \text{ and } k = 5,$$

and $\quad P(\mu - k\sigma \leq X \leq \mu + k\sigma) \geq 1 - 1/(5)^2$
$$\geq 1 - 1/25$$
$$\geq 24/25, \text{ or at least } 0.96.$$

23. Using Chebychev's inequality we have

$$P(\mu - k\sigma \leq X \leq \mu + k\sigma) \geq 1 - 1/k^2.$$

Here $k = 2$, so
$$P(\mu - k\sigma \leq X \leq \mu + k\sigma) \geq 1 - 1/2^2 = 3/4.$$
This means that at least 75 percent of the values are expected to lie between $\mu - 2\sigma$ and $\mu + 2\sigma$.

25. Using Chebychev's inequality we have
$$P(\mu - k\sigma \leq X \leq \mu + k\sigma) \geq 1 - 1/k^2.$$
Here, $\mu - k\sigma = 24 - k(3) = 20$, and $k = 4/3$.
So $P(\mu - k\sigma \leq X \leq \mu + k\sigma) \geq 1 - 1/(4/3)^2 \geq 1 - 9/16 = 7/16$.
or at least 0.4375.

27. Here $\mu = 5$ and $\sigma = 0.02$. Next, we require that $c = k\sigma$, or $k = \dfrac{c}{0.02}$. Next, solve
$$0.96 = 1 - \left(\tfrac{0.02}{c}\right)^2$$

$$0.04 = \frac{0.0004}{c^2} = 0.01$$

$$c^2 = \frac{0.0004}{0.04} = 0.01 \quad \text{and} \quad c = 0.1.$$
We conclude that for $P(5 - c \leq X \leq 5 + c) \geq 0.96$. We require that $c \geq 0.1$.

29. True. If $k \leq 1$, then $\dfrac{1}{k} \geq 1$ and so $1 - \dfrac{1}{k^2} \leq 0$ and the inequality (9) becomes trivial.

USING TECHNOLOGY EXERCISES, page 479

1. a.

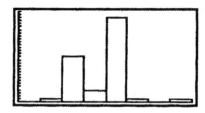

b. $\mu = 4$ and $\sigma = 1.40$

3. a.

b. $\mu = 17.34$ and $\sigma = 1.11$

5.   a. Let X denote the random variable that gives the weight of a carton of sugar.
      b. The probability distribution for the random variable X is

x	4.96	4.97	4.98	4.99	5.00	5.01	5.02	5.03
$P(X=x)$	$\dfrac{3}{30}$	$\dfrac{4}{30}$	$\dfrac{4}{30}$	$\dfrac{1}{30}$	$\dfrac{1}{30}$	$\dfrac{5}{30}$	$\dfrac{3}{30}$	$\dfrac{3}{30}$

x	5.04	5.05	5.06
$P(X=x)$	$\dfrac{4}{30}$	$\dfrac{1}{30}$	$\dfrac{1}{30}$

$\mu = 5.00467 \approx 5.00; \quad V(X) = 0.0009; \quad \sigma = \sqrt{0.0009} = 0.03$

7. a.   b. $\mu = 65.875$ and $\sigma = 1.73$.

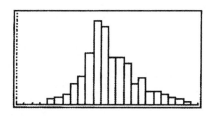

EXERCISES 8.4, page 504

1.   Yes. The number of trials is fixed, there are two outcomes of the experiment, the probability in each trial is fixed $(p = \frac{1}{6})$, and the trials are independent of each other.

3.   No. There are more than 2 outcomes in each trial.

5.   No. There are more than 2 outcomes in each trial and the probability of success (an accident) in each trial is not the same.

7. $C(4,2)(\frac{1}{2})^2(\frac{2}{3})^2 = \dfrac{4!}{2!2!}(\frac{4}{81}) \approx 0.296$.

9. $C(5,3)(0.2)^3(0.8)^2 = (\frac{5!}{2!3!})(0.2)^3(0.8)^2 \approx 0.0512.$

11.   The required probability is given by
$$P(X = 0) = C(5,0)(\tfrac{1}{3})^0(\tfrac{2}{3})^5 \approx 0.132.$$

13.   The required probability is given by
$$P(X \geq 3) = C(6,3)(\tfrac{1}{2})^3(\tfrac{1}{2})^{6-3} + C(6,4)(\tfrac{1}{2})^4(\tfrac{1}{2})^{6-4} + C(6,5)(\tfrac{1}{2})^5(\tfrac{1}{2})^{6-5}$$
$$+ C(6,6)(\tfrac{1}{2})^6(\tfrac{1}{2})^{6-6}$$
$$= \tfrac{6!}{3!3!}(\tfrac{1}{2})^6 + \tfrac{6!}{4!2!}(\tfrac{1}{2})^6 + \tfrac{6!}{5!1!}(\tfrac{1}{2})^6 + \tfrac{6!}{6!0!}(\tfrac{1}{2})^6$$
$$= \tfrac{1}{64}(20+15+6+1) = \tfrac{21}{32}.$$

15.   The probability of no failures, or, equivalently, the probability of five successes is
$$P(X = 5) = C(5,5)(\tfrac{1}{3})^5(\tfrac{2}{3})^{5-5} = \tfrac{1}{243} \approx 0.00412.$$

17.   Here $n = 4$, and $p = 1/6$. Then
$$P(X = 2) = C(4,2)(\tfrac{1}{6})^2(\tfrac{5}{6})^2 = (\tfrac{25}{216}) \approx 0.116.$$

19.   Here $n = 5$, $p = 0.4$, and therefore, $q = 1 - 0.4 = 0.6$.
a. Using Table 2 in the Appendix, we see that
     $P(X = 0) = 0.078$, $P(X = 1) = 0.259$, $P(X = 2) = 0.346$,
     $P(X = 3) = 0.230$, $P(X = 4) = 0.077$, $P(X = 5) = 0.010$.
b.

x	0	1	2	3	4	5
$P(X=x)$	0.078	0.259	0.346	0.230	0.077	0.010

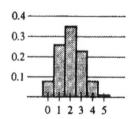

c. $\mu = np = 5(0.4) = 2$; $\sigma = \sqrt{npq} = \sqrt{5(0.4)(0.6)} = \sqrt{1.2} = 1.095.$

21. Here $1 - p = 1/50$ or $p = 49/50$. So the probability of obtaining 49 or 50 nondefective fuses is

$$P(X = 49) + P(X = 50) = C(50,49)(\tfrac{49}{50})^{49}(\tfrac{1}{50}) + C(50,50)(\tfrac{49}{50})^{50}(\tfrac{1}{50})^{0} \approx 0.74.$$

This is also the probability of at most one defective fuse. So the inference is incorrect.

23. The required probability is given by

$$P(X = 6) = C(6,6)(\tfrac{1}{4})^{6}(\tfrac{3}{4})^{0} = 0.0002.$$

25. a. The probability that six or more people stated a preference for brand A is

$$P(X \geq 6) = C(10,6)(.6)^{6}(.4)^{4} + C(10,7)(.6)^{7}(.4)^{3}$$
$$+ C(10,8)(.6)^{8}(.4)^{2} + C(10,9)(.6)^{9}(.4)^{1}$$
$$+ C(10,10)(.6)^{10}(.4)^{0}$$

$$\approx 0.251 + 0.215 + 0.121 + 0.040 + 0.006 = 0.633.$$

(Use Table 2 in Appendix C.)

b. The required probability is $1 - 0.633 = 0.367$.

27. a. Let X denote the number of new buildings that are in violation of the building code. then $p = \tfrac{1}{3}$ and $p = \tfrac{2}{3}$. Therefore, the probability that the first 3 new buildings will pass inspection and the remaining 2 will fail the inspection is

$$\left(\tfrac{2}{3}\right)^{2}\left(\tfrac{1}{3}\right)^{2} = \frac{2^{3}}{3^{5}},$$

or approximately 0.0329.

b. The probability that just 3 of the new buildings will pass inspection (and so 2 will fail the inspection) is

$$C(5,2)\left(\tfrac{1}{3}\right)^{2}\left(\tfrac{2}{3}\right)^{3} = \frac{5!}{3!2!} \cdot \frac{2^{3}}{3^{5}},$$

or approximately 0.3292.

29. This is a binomial experiment with $n = 9$, $p = 1/3$, and $q = 2/3$.
a. The probability is given by

$$P(X = 3) = C(9,3)(\tfrac{1}{3})^{3}(\tfrac{2}{3})^{6} = (\tfrac{9!}{6!3!})(\tfrac{1}{3})^{3}(\tfrac{2}{3})^{6} \approx 0.273.$$

b. The probability is given by
$$P(X=0) + P(X=1) + P(X=2) + P(X=3)$$
$$= C(9,0)(\tfrac{1}{3})^0(\tfrac{2}{3})^9 + C(9,1)(\tfrac{1}{3})(\tfrac{2}{3})^8 + C(9,2)(\tfrac{1}{3})^2(\tfrac{2}{3})^7 + C(9,3)(\tfrac{1}{3})^3(\tfrac{2}{3})^6$$
$$= 0.026 + 0.117 + 0.234 + 0.273 \approx 0.650.$$

31. In order to obtain a score of at least 90 percent the student needs to answer 3 or 4 of the remaining questions correctly. The probability of doing this is
$$P(X \geq 3) = C(4,3)(0.5)^3(0.5) + C(4,4)(0.5)^4(0.5)^0 = 0.3125.$$

33. Here $1 - p = 0.015$ and $p = 0.985$. The probability that the sample will be accepted, or, equivalently, that the sample contains all six nondefective cartridges is given by
$$P(X = 6) = C(6,6)(0.985)^6(0.015)^0 \approx 0.913.$$

35. This is a binomial experiment with $n = 4$, $p = 0.001$, and $q = 0.999$.
a. The probability that exactly one engine will fail is given by
$$P(X = 1) = C(4,1)(0.001)(0.999)^3 = 4(0.001)(0.999)^3 = 0.003988.$$
b. The probability that exactly two engines will fail is given by
$$P(X = 2) = C(4,2)(0.001)^2(0.999)^2 = 6(0.001)^2(0.999)^2$$
$$= 0.000006.$$
c. The probability that more than two engines will fail is given by
$$P(X > 2) = P(X = 3) + P(X = 4)$$
$$= C(4,3)(0.001)^3(0.999) + C(4,4)(0.001)^4(0.999)^0$$
$$= 3.996 \times 10^{-9} + 1 \times 10^{-12} \approx 3.997 \times 10^{-9}.$$

37. The required probability is
$$P(X \leq 1) = P(X = 0) + P(X = 1)$$
$$= C(20,0)(0.1)^0(0.9)^{20} + C(20,1)(0.1)^1(0.9)^{19} \approx 0.3917.$$

39. Take $p = 1/2$. The probability of obtaining no heads in n tosses is
$$P(X = n) = C(n,n)(\tfrac{1}{2})^n(\tfrac{1}{2})^0 = (\tfrac{1}{2})^n.$$

The probability of obtaining at least one head is $1 - (\tfrac{1}{2})^n$. We want this to exceed 0.99. Thus,
$$1 - \left(\frac{1}{2}\right)^n \geq 0.99$$
$$\frac{1}{2^n} \geq 0.01$$
$$2^n \geq 100$$

or $\qquad n \geq \dfrac{\ln 100}{\ln 2} \approx 6.64$

So one must toss the coin at least 7 times.

41.   a. The expected number of students who will graduate within four years is
$$\mu = np = (0.6)(2000) = 1200.$$

b. The standard deviation of the number of students who will graduate within four years is
$$\sigma = \sqrt{npq} = \sqrt{(2000)(0.6)(0.4)} \approx 21.91.$$

43.   True.
$$\begin{aligned}
P(X = 1 \text{ or } 2) &= P(X = 1) + P(X = 2) \\
&= C(3,1)p^1 q^2 + C(3,2)(p)^2 q \\
&= \frac{3!}{2!1!} pq^2 + \frac{3!}{1!2!} p^2 q \\
&= 3pq^2 + 3p^2 q = 3pq(q + p) \\
&= 3pq(1) = 3pq
\end{aligned}$$

EXERCISES 8.5, page 515

1. $P(Z < 1.45) = 0.9265.$

3. $P(Z < -1.75) = 0.0401.$

5. $P(-1.32 < Z < 1.74) = P(Z < 1.74) - P(Z < -1.32)$
$\qquad\qquad\qquad\qquad\qquad = 0.9591 - 0.0934 = 0.8657.$

7. $P(Z < 1.37) = 0.9147.$

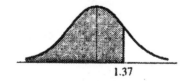

1.37

9. $P(Z < -0.65) = 0.2578$.

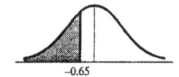

11. $P(Z > -1.25) = 1 - P(Z < -1.25) = 1 - 0.1056 = 0.8944$

13. $P(0.68 < Z < 2.02) = P(Z < 2.02) - P(Z < 0.68)$
$$= 0.9783 - 0.7517 = 0.2266.$$

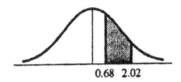

15. a. Referring to Table 3, we see that $P(Z < z) = 0.8907$ implies that $z = 1.23$.

 b. Referring to Table 3, we see that $P(Z < z) = 0.2090$ implies that $z = -0.81$.

17. a. $P(Z > -z) = 1 - P(Z < -z) = 1 - 0.9713 = 0.0287$ implies $z = 1.9$.
 b. $P(Z < -z) = 0.9713$ implies that $z = -1.9$.

19. a. $P(X < 60) = P(Z < \dfrac{60 - 50}{5}) = P(Z < 2) = 0.9772$.

 b. $P(X > 43) = P(Z > \dfrac{43 - 50}{5}) = P(Z > -1.4) = P(Z < 1.4) = 0.9192$.

 c. $P(46 < X < 58) = P(\dfrac{46 - 50}{5} < Z < \dfrac{58 - 50}{5}) = P(-0.8 < Z < 1.6)$
$$= P(Z < 1.6) - P(Z < -0.8)$$
$$= 0.9452 - 0.2119 = 0.7333.$$

EXERCISES 8.6, page 525

1. $\mu = 20$ and $\sigma = 2.6$.

    a. $P(X > 22) = P(Z > \dfrac{22 - 20}{2.6}) = P(Z > 0.77) = P(Z < -0.77) = 0.2206.$

    b. $P(X < 18) = P(Z < \dfrac{18 - 20}{2.6}) = P(Z < -0.77) = 0.2206.$

    c. $P(19 < X < 21) = P(\dfrac{19 - 20}{2.6} < Z < \dfrac{21 - 20}{2.6}) = P(-0.39 < Z < 0.39)$

               $= P(Z < 0.39) - P(Z < -0.39)$

               $= 0.6517 - 0.3483 = 0.3034.$

3. $\mu = 750$ and $\sigma = 75$.

    a. $P(X > 900) = P(Z > \dfrac{900 - 750}{75}) = P(Z > 2) = P(Z < -2) = 0.0228.$

    b. $P(X < 600) = P(Z < \dfrac{600 - 750}{75}) = P(Z < -2) = 0.0228.$

    c. $P(750 < X < 900) = P(Z < \dfrac{750 - 750}{75} < Z < \dfrac{900 - 750}{75})$

                     $= P(0 < Z < 2) = P(Z < 2) - P(Z < 0)$

                     $= 0.9772 - 0.5000 = 0.4772.$

    d. $P(600 < X < 800) = P(\dfrac{600 - 750}{75} < Z < \dfrac{800 - 750}{75})$

                     $= P(-2 < Z < .667) = P(Z < .667) - P(Z < -2)$

                     $= 0.7486 - 0.0228 = 0.7258.$

5. $\mu = 100$ and $\sigma = 15$.

    a. $P(X > 140) = P(Z > \dfrac{140 - 100}{15}) = P(Z > 2.667) = P(Z < -2.667)$

          $= 0.0038.$

    b. $P(X > 120) = P(Z > \dfrac{120 - 100}{15}) = P(Z > 1.33) = P(Z < -1.33)$

          $= 0.0918.$

    c. $P(100 < X < 120) = P(\dfrac{100 - 100}{15} < Z < \dfrac{120 - 100}{15}) = P(0 < Z < 1.333)$

$$= P(Z<0) - P(Z<1.333)$$

$$= 0.9082 - 0.5000 = 0.4082.$$

d. $P(X<90) = P(Z < \dfrac{90-100}{15}) = P(Z<-0.667) = 0.2514.$

7.   Here $\mu = 475$ and $\sigma = 50$.

$$P(450<X<550) = P(\dfrac{450-475}{50}<Z<\dfrac{550-475}{50}) = P(-0.5<Z<1.5)$$
$$= P(Z<1.5) - P(Z<-0.5)$$
$$= 0.9332 - 0.3085 = 0.6247.$$

9.   Here $\mu = 22$ and $\sigma = 4$.

$$P(X<12) = P(Z<\dfrac{12-22}{4}) = P(Z<-2.5) = 0.0062, \text{ or } 0.62 \text{ percent.}$$

11. $\mu = 70$ and $\sigma = 10$.
To find the cut-off point for an A, we solve $P(Y<y) = 0.85$ for y. Now

$$P(Y<y) = P\left(Z<\dfrac{y-70}{10}\right) = 0.85 \text{ implies } \dfrac{y-72}{10} = 1.04$$

or $y = 80.4 \approx 80$.

For a B: $P(Y<y) = P\left(Z<\dfrac{y-70}{10}\right) = 0.60 \text{ implies } \dfrac{y-70}{10} = 0.25, \text{ or } y \approx 73.$

For a C: $P\left(Z \le \dfrac{y-70}{10}\right) = 0.2 \text{ implies } \dfrac{y-70}{10} = -0.84 \text{ or } y \approx 62.$

For a D: $P\left(Z<\dfrac{y-70}{10}\right) = 0.05 \text{ implies } \dfrac{y-70}{10} = -1.65, \text{ or } y \approx 54.$

13.   Let X denote the number of heads in 25 tosses of the coin. Then X is a binomial random variable. Also, $n = 25, p = 0.4,$ and $q = 0.6$. So
$$\mu = (25)(0.4) = 10$$
$$\sigma = \sqrt{(25)(0.4)(0.6)} \approx 2.45.$$
Approximating the binomial distribution by a normal distribution with a mean of 10 and a standard deviation of 2.45, we find upon letting Y denote the associated normal random variable,

a. $P(X < 10) \approx P(Y < 9.5)$
$$= P\left(Z < \frac{9.5 - 10}{2.45}\right) = P(Z < -0.20) = 0.4207.$$

b. $P(10 \le X \le 12) \approx P(9.5 < Y < 12.5)$
$$= P\left(\frac{9.5 - 10}{2.45} < Z < \frac{12.5 - 10}{2.45}\right) = P(Z < 1.02) - P(Z < -0.20)$$
$$= P(Z < 1.02) - P(Z < -0.20) = 0.8461 - 0.4207 = 0.4254.$$

c. $P(X > 15) \approx P(Y \ge 15)$
$$= P(Z > \frac{15.5 - 10}{2.45}) = P(Z > 2.25) = P(Z < -2.25) = 0.0122.$$

15. Let X denote the number of times the marksman hits his target. Then X has a binomial distribution with $n = 30$, $p = 0.6$ and $q = 0.4$. Therefore,
$$\mu = (30)(0.6) = 18, \quad \sigma = \sqrt{(30)(0.6)(0.4)} = 2.68.$$

a. $P(X \ge 20) \approx P(Y \ge 19.5)$
$$= P\left(Z > \frac{19.5 - 18}{2.68}\right) = P(Z > 0.56) = P(Z < -0.56) = 0.2877.$$

b. $P(X < 10) \approx P(Y < 9.5) = P\left(Z < \frac{9.5 - 18}{2.68}\right) = P(Z < -3.17) = 0.0008.$

c. $P(15 \le X \le 20) \approx P(14.5 < Y < 20.5) = P\left(\frac{14.5 - 18}{2.68} < Z < \frac{20.5 - 18}{2.68}\right)$
$$= P(Z < 0.93) - P(Z < -1.31)$$
$$= 0.8238 - 0.0951 = 0.7287.$$

17. Let X denote the number of "seconds." Then X has a binomial distribution with $n = 200$, $p = 0.03$, and $q = 0.97$. Then
$$\mu = (200)(0.03) = 6$$
$$\sigma = \sqrt{(200)(0.03)(0.97)} \approx 2.41,$$
and $P(X < 10) \approx P(Y < 9.5) = P(Z < \frac{9.5 - 6}{2.41}) = P(Z < 1.45) = 0.9265.$

19. a. Let X denote the number of mice that recovered from the disease. Then X has a binomial distribution with $n = 50$, $p = 0.5$, and $q = 0.5$, so
$$\mu = (50)(0.5) = 25$$

$$\sigma = \sqrt{(50)(0.5)(0.5)} \approx 3.54,$$

Approximating the binomial distribution by a normal distribution with a mean of 25 and a standard deviation of 3.54, we find that the probability that 35 or more of the mice would recover from the disease without benefit of the drug is

$$P(X \geq 35) \approx P(Y > 34.5)$$
$$= P(Z > \frac{34.5 - 25}{3.54})$$
$$= P(Z > 2.68) = P(Z < -2.68) = 0.0037.$$

b. The drug is very effective.

21. Let n denote the number of reservations the company should accept. Then we need to find

$$P(X \geq 2000) \approx P(Y > 1999.5) = 0.01$$

or equivalently,

$$P(Z \geq \frac{1999.5 - np}{\sqrt{npq}}) = 0.01 \qquad [\text{Here } p = 0.92 \text{ and } q = 0.08.]$$

Using Table 3 in Appendix C, we find

$$\frac{1999.5 - 0.92n}{\sqrt{0.0736n}} = -2.33$$

$$(1999.5 - 0.92n)^2 = (-2.33)^2(0.0736n)$$

$$3,998,000.25 - 3679.08n + 0.8464n^2 = 0.39956704n,$$

or $\qquad 0.8464n^2 - 3679.479567n + 3,998,000.25 = 0$.
Using the quadratic formula, we obtain

$$n = \frac{3679.479567 \pm \sqrt{2940.2375}}{1.6928} \approx 2142,$$

or 2142. [You can verify that 2206 is not a root of the original equation (before squaring).] Therefore, the company should accept no more than 2142 reservations.

CHAPTER 8, REVIEW EXERCISES, page 528

1. a. $S = \{WWW, WWB, WBW, WBB, BWW, BWB, BBW, BBB\}$

b.

Outcome	WWW	WWB	WBW	WBB
Value	0	1	1	2

Outcome	BWW	BWB	BBW	BBB
Value	1	2	2	3

c.

x	0	1	2	3
$P(X=x)$	$\frac{1}{35}$	$\frac{12}{35}$	$\frac{18}{35}$	$\frac{4}{35}$

d.

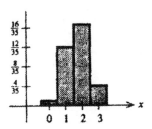

3. a. $P(1 \leq X \leq 4) = 0.1 + 0.2 + 0.3 + 0.2 = 0.8.$

   b. $\mu = 0(0.1) + 1(0.1) + 2(0.2) + 3(0.3) + 4(0.2) + 5(0.1) = 2.7.$

$$V(X) = 0.1(0 - 2.7)^2 + 0.1(1 - 2.7)^2 + 0.2(2 - 2.7)^2 + 0.3(3 - 2.7)^2$$
$$+ 0.2(4 - 2.7)^2 + 0.1(5 - 2.7)^2$$
$$= 2.01$$

$$\sigma = \sqrt{2.01} \approx 1.418.$$

5. $P(Z < 0.5) = 0.6915.$

7. $P(-0.75 < Z < 0.5) = P(Z < 0.5) - P(Z < -0.75)$
   $= 0.6915 - 0.2266 = 0.4649.$

-0.75 0 0.5

9. If $P(Z < z) = 0.9922$, then $z = 2.42$.

11. If $P(Z > z) = 0.9788$, then $P(Z < -z) = 0.9788$, and $-z = 2.03$,
    or $z = -2.03$.

13. $P(X < 11) = P\left(Z < \dfrac{11-10}{2}\right) = P(Z < 0.5) = 0.6915.$

15. $P(7 < X < 9) = P\left(\dfrac{7-10}{2} < Z < \dfrac{9-10}{2}\right) = P(-1.5 < Z < -0.5)$
    $= P(Z < -0.5) - P(Z < -1.5)$
    $= 0.3085 - 0.0668 = 0.2417.$

17. This is a binomial experiment with $p = 0.7$, and so $q = 0.3$. The probability that he
    will get exactly two strikes in four attempts is given by
    $P(X = 2) = C(4,2)(0.7)^2(0.3)^2 \approx 0.2646.$
    The probability that he will get at least two strikes in four attempts is given by
    $P(X = 2) + P(X = 3) + P(X = 4)$
    $= C(4,2)(0.7)^2(0.3)^2 + C(4,3)(0.7)^3(0.3) + C(4,4)(0.7)^4(0.3)^0$
    $= 0.2646 + 0.4116 + 0.2401 \approx 0.9163.$

19. Here $\mu = 64.5$ and $\sigma = 2.5$. Next,
    $64.5 - 2.5k = 59.5$ and $64.5 + 2.5k = 69.5$
    and $k = 2$. Therefore, the required probability is given by
    $P(59.5 \le X \le 69.5) \ge 1 - \dfrac{1}{2^2} = 0.75.$

21. Let the random variable X be the number of people for whom the drug is effective.

Then
$$\mu = (0.15)(800) = 120 \text{ and } \sigma = \sqrt{(800)(0.15)(0.85)} = \sqrt{102} \approx 10.1.$$

23. Here $\mu = (0.6)(100) = 60$ and $\sigma = \sqrt{100(0.6)(0.4)} = 4.899$.

Then $P(X > 50) \approx P(Y > 50.5) = P\left(Z > \dfrac{50.5 - 60}{4.899}\right) = P(Z > -1.94)$

$$= P(Z < 1.94) = 0.9738.$$

CHAPTER 9

EXERCISES 9.1, page 53

1. Yes. All entries are nonnegative and the sum of the entries in each column is equal to 1.

3. Yes.

5. No. The sum of the entries of the third column is not 1.

7. Yes.

9. No. It is not a square ($n \times n$) matrix.

11. a. The conditional probability that the outcome state 1 will occur given that the outcome state 1 has occurred is 0.3.
    b. 0.7
    c. We compute $X_1 = TX_0 = \begin{bmatrix} .3 & .6 \\ .7 & .4 \end{bmatrix} \begin{bmatrix} .4 \\ .6 \end{bmatrix} = \begin{bmatrix} .48 \\ .52 \end{bmatrix}$.

13. We compute $TX_0 = \begin{bmatrix} .6 & .2 \\ .4 & .8 \end{bmatrix} \begin{bmatrix} .5 \\ .5 \end{bmatrix} = \begin{bmatrix} .4 \\ .6 \end{bmatrix}$

    Thus, after 1 stage of the experiment, the probability of state 1 occurring is 0.4 and the probability of state 2 occurring is 0.6. The tree diagram describing this process follows.

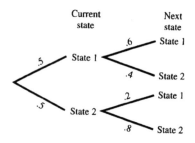

    Using this diagram, we see that the probabilities of state 1 and state 2 occurring in the next stage of the experiment are given by

$$P(S_1) = (0.5)(0.6) + (0.5)(0.2) = 0.4$$
$$P(S_2) = (0.5)(0.4) + (0.5)(0.8) = 0.6$$

Observe that these probabilities are precisely those represented in the probability distribution vector TX_0.

15. $X_1 = TX_0 = \begin{bmatrix} .4 & .8 \\ .6 & .2 \end{bmatrix}\begin{bmatrix} .6 \\ .4 \end{bmatrix} = \begin{bmatrix} .56 \\ .44 \end{bmatrix}.$

$X_2 = TX_1 = \begin{bmatrix} .4 & .6 \\ .8 & .2 \end{bmatrix}\begin{bmatrix} .56 \\ .44 \end{bmatrix} = \begin{bmatrix} .576 \\ .424 \end{bmatrix}.$

17. $X_1 = TX_0 = \begin{bmatrix} \frac{1}{4} & \frac{1}{4} & \frac{1}{2} \\ \frac{1}{4} & \frac{1}{2} & \frac{1}{2} \\ \frac{1}{2} & \frac{1}{4} & 0 \end{bmatrix}\begin{bmatrix} \frac{1}{4} \\ \frac{1}{2} \\ \frac{1}{4} \end{bmatrix} = \begin{bmatrix} \frac{5}{16} \\ \frac{7}{16} \\ \frac{1}{4} \end{bmatrix}$

$X_2 = TX_0 = \begin{bmatrix} \frac{1}{4} & \frac{1}{4} & \frac{1}{2} \\ \frac{1}{4} & \frac{1}{2} & \frac{1}{2} \\ \frac{1}{2} & \frac{1}{4} & 0 \end{bmatrix}\begin{bmatrix} \frac{5}{16} \\ \frac{7}{16} \\ \frac{1}{4} \end{bmatrix} = \begin{bmatrix} \frac{5}{16} \\ \frac{27}{64} \\ \frac{17}{64} \end{bmatrix}.$

19. a.

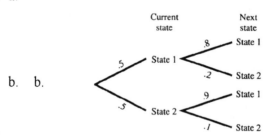

b. b.

c. $T = \begin{array}{c} \\ L \\ R \end{array}\begin{array}{c} L \quad R \\ \begin{bmatrix} .8 & .9 \\ .2 & .1 \end{bmatrix} \end{array}$

c. $X_0 = \begin{array}{c} L \\ R \end{array}\begin{bmatrix} .5 \\ .5 \end{bmatrix}$

d. $X_1 = \begin{array}{c} \\ L \\ R \end{array}\begin{array}{c} L \quad R \\ \begin{bmatrix} .8 & .9 \\ .2 & .1 \end{bmatrix} \end{array}\begin{bmatrix} .5 \\ .5 \end{bmatrix} = \begin{array}{c} L \\ R \end{array}\begin{bmatrix} .85 \\ .15 \end{bmatrix}$

21. a.

$$X_1 = TX_0 = \begin{matrix} & R & D \\ R \\ D \end{matrix}\begin{bmatrix} .7 & .2 \\ .3 & .8 \end{bmatrix}\begin{bmatrix} .6 \\ .4 \end{bmatrix} = \begin{matrix} R \\ D \end{matrix}\begin{bmatrix} .5 \\ .5 \end{bmatrix}$$

so if the election were held now, it would be a tie.

b.

$$X_1 = TX_0 = \begin{matrix} & R & D \\ R \\ D \end{matrix}\begin{bmatrix} .7 & .2 \\ .3 & .8 \end{bmatrix}\begin{bmatrix} .5 \\ .5 \end{bmatrix} = \begin{matrix} R \\ D \end{matrix}\begin{bmatrix} .45 \\ .55 \end{bmatrix}$$

so the Democratic candidate would win.

23. After one pickup and discharge the distribution will be

$$X_1 = TX_0 = \begin{bmatrix} .6 & .4 & .3 \\ .3 & .3 & .3 \\ .1 & .3 & .4 \end{bmatrix}\begin{bmatrix} .6 \\ .2 \\ .2 \end{bmatrix} = \begin{bmatrix} .5 \\ .3 \\ .2 \end{bmatrix}$$

or 50% will be in Zone I, 30% will be in Zone II, and 20 percent will be in Zone III.

25. The expected distribution is given by

$$X_1 = TX_0 = \begin{bmatrix} .80 & .10 & .05 \\ .10 & .75 & .05 \\ .10 & .15 & .90 \end{bmatrix}\begin{bmatrix} .4 \\ .4 \\ .2 \end{bmatrix} = \begin{bmatrix} .37 \\ .35 \\ .28 \end{bmatrix}$$

and we conclude that at the beginning of the second quarter the university Bookstore will have 37 percent of the market, the Campus Bookstore will have 35 percent, and the Book Mart will have 28 percent of the market.
Similarly,

$$X_2 = TX_2 = \begin{bmatrix} .80 & .10 & .05 \\ .10 & .75 & .05 \\ .10 & .15 & .90 \end{bmatrix}\begin{bmatrix} .37 \\ .35 \\ .28 \end{bmatrix} = \begin{bmatrix} .345 \\ .3135 \\ .3415 \end{bmatrix}$$

implies that the University Bookstore will have 34.5% of the market, the Campus Bookstore will have 31.35% of the market, and the Book Mart will have 34.15% of the market at the beginning of the third quarter.

27. $X_2 = TX_0 = \begin{bmatrix} .80 & .10 & .20 & .10 \\ .10 & .70 & .10 & .05 \\ .05 & .10 & .60 & .05 \\ .05 & .10 & .10 & .80 \end{bmatrix} \begin{bmatrix} .3 \\ .3 \\ .2 \\ .2 \end{bmatrix} = \begin{bmatrix} .33 \\ .17 \\ .175 \\ .225 \end{bmatrix}$

Similarly

$$X_2 = TX_1 = \begin{bmatrix} .3485 \\ .25075 \\ .15975 \\ .241 \end{bmatrix} \quad \text{and} \quad X_3 = TX_2 = \begin{bmatrix} .3599 \\ .2384 \\ .1504 \\ .2513 \end{bmatrix}$$

Assuming that the present trend continues, 36% of the students in their senior year will major in business, 23.8% will major in the humanities, 15% will major in education, and 25.1% will major in the natural sciences.

29. False. In a Markov Chain, an outcome depends only on the preceding stage.

USING TECHNOLOGY EXERCISES 9.1, page 543

1. $X_5 = \begin{bmatrix} .204489 \\ .131869 \\ .261028 \\ .186814 \\ .2158 \end{bmatrix}$

3. Manufacturer A will have 23.95% of the market, Manufacturer B will have 49.71% of the market share, and manufacturer C will have 26.34 percent of the market share.

EXERCISES 9.2, page 552

1. Since all entires in the matrix are positive, it is regular.

3. $T^2 = \begin{bmatrix} 1 & .8 \\ 0 & .2 \end{bmatrix} \begin{bmatrix} 1 & .8 \\ 0 & .2 \end{bmatrix} = \begin{bmatrix} 1 & .96 \\ 0 & .04 \end{bmatrix};$ $T^3 = \begin{bmatrix} 1 & .96 \\ 0 & .04 \end{bmatrix} \begin{bmatrix} 1 & .8 \\ 0 & .2 \end{bmatrix} = \begin{bmatrix} 1 & .992 \\ 0 & .008 \end{bmatrix}$

and we see that the a_{12} entry will always be zero, so T is not regular.

5. $T^2 = \begin{bmatrix} \frac{1}{2} & \frac{3}{4} & 0 \\ \frac{1}{2} & 0 & \frac{1}{2} \\ 0 & \frac{1}{4} & \frac{1}{2} \end{bmatrix} \begin{bmatrix} \frac{1}{2} & \frac{3}{4} & 0 \\ \frac{1}{2} & 0 & \frac{1}{2} \\ 0 & \frac{1}{4} & \frac{1}{2} \end{bmatrix} = \begin{bmatrix} \frac{5}{8} & \frac{3}{8} & \frac{3}{8} \\ \frac{1}{4} & \frac{1}{2} & \frac{1}{4} \\ \frac{1}{8} & \frac{1}{8} & \frac{3}{8} \end{bmatrix}$

and so this matrix is regular.

7. $T^2 = \begin{bmatrix} .7 & .2 & .3 \\ .3 & .8 & .3 \\ 0 & 0 & .4 \end{bmatrix} \begin{bmatrix} .7 & .2 & .3 \\ .3 & .8 & .3 \\ 0 & 0 & .4 \end{bmatrix} = \begin{bmatrix} .55 & .3 & .39 \\ .45 & .7 & .45 \\ 0 & 0 & .16 \end{bmatrix}$

and so forth.

Continuing, we see that $T^3, T^4, \ldots$, will have the a_{13} and a_{23} entries equal to zero and T is not regular.

9. We solve the matrix equation

$$\begin{bmatrix} \frac{1}{3} & \frac{1}{4} \\ \frac{2}{3} & \frac{3}{4} \end{bmatrix} \begin{bmatrix} x \\ y \end{bmatrix} = \begin{bmatrix} x \\ y \end{bmatrix}$$

or equivalently, the system of equations
$$\frac{1}{3}x + \frac{1}{4}y = x$$
$$\frac{2}{3}x + \frac{3}{4}y = y$$
$$x + y = 1.$$

Solving this system of equations, we find the required vector to be

$$\begin{bmatrix} \frac{3}{11} \\ \frac{8}{11} \end{bmatrix}.$$

11. We have $TX = X$, that is,
$$\begin{bmatrix} .5 & .2 \\ .5 & .8 \end{bmatrix} \begin{bmatrix} x \\ y \end{bmatrix} = \begin{bmatrix} x \\ y \end{bmatrix}$$
or equivalently, the sytem of equations
$$.5x + .2y = x$$
$$.5x + .8y = y.$$

These two equations are equivalent to the single equation $0.5x - 0.2y = 0$.

We must also have $x + y = 1$. So we have the system

$$.5x - .2y = x$$
$$x + \ y = 1$$

The second equation gives $y = 1 - x$, which when substituted into the first equation yields,

$$0.5x - 0.2(1 - x) = 0, \ 0.7x - 0.2 = 0, \ \text{or} \ x = 2/7.$$

Therefore, $y = 5/7$ and the steady-state distribution vector is

$$\begin{bmatrix} \frac{2}{7} \\ \frac{5}{7} \end{bmatrix}.$$

13. We solve the system

$$\begin{bmatrix} 0 & \frac{1}{8} & 1 \\ 1 & \frac{5}{8} & 0 \\ 1 & \frac{1}{4} & 0 \end{bmatrix} \begin{bmatrix} x \\ y \\ z \end{bmatrix} = \begin{bmatrix} x \\ y \\ z \end{bmatrix}$$

together with the equation $x + y + z = 1$; that is, the system

$$-x + \tfrac{1}{8}y + z = 0$$
$$x - \tfrac{3}{8}y \qquad = 0$$
$$\tfrac{1}{4}y - z = 0$$
$$x + y + z = 1.$$

Using the Gauss-Jordan method, we find that the required steady-state vector is

$$\begin{bmatrix} \frac{3}{13} \\ \frac{8}{13} \\ \frac{2}{13} \end{bmatrix}.$$

15. We solve the system

$$\begin{bmatrix} .2 & 0 & .3 \\ 0 & .6 & .4 \\ .8 & .4 & .3 \end{bmatrix} \begin{bmatrix} x \\ y \\ z \end{bmatrix} = \begin{bmatrix} x \\ y \\ z \end{bmatrix}$$

together with the equation $x + y + z = 1$, or equivalently, the system

$$-0.8x \qquad +0.3z = 0$$
$$-0.4y + 0.4z = 0$$
$$0.8x + 0.4y - 0.7z = 0$$
$$x \quad +y \quad +z = 1.$$

Using the Gauss-Jordan method, we find that the required steady-state vector is

$$\begin{bmatrix} \frac{3}{19} \\ \frac{8}{19} \\ \frac{8}{19} \end{bmatrix}.$$

17. a. We want to solve
$$\begin{bmatrix} .8 & .9 \\ .2 & .1 \end{bmatrix} \begin{bmatrix} x \\ y \end{bmatrix} = \begin{bmatrix} x \\ y \end{bmatrix},$$

or equivalently
$$.8x + .9y = x$$
$$.2x + .1y = y$$
$$x + y = 1$$

Solving this system, we find that the required steady-state vector is $\begin{bmatrix} \frac{2}{11} \\ \frac{9}{11} \end{bmatrix}$

and conclude that in the long run, the mouse will turn left 81.8% of the time.

19. We compute

$$X_1 = \begin{bmatrix} .72 & .12 \\ .28 & .88 \end{bmatrix} \begin{bmatrix} .48 \\ .52 \end{bmatrix} = \begin{bmatrix} .408 \\ .592 \end{bmatrix},$$

and conclude that, ten years from now, there will be 40.8 percent 1-wage-earners and 59.2% 2-wage earners.

b. We solve the system
$$\begin{bmatrix} .72 & .12 \\ .28 & .88 \end{bmatrix} \begin{bmatrix} x \\ y \end{bmatrix} = \begin{bmatrix} x \\ y \end{bmatrix}$$
together with the equation $x + y = 1$.

$$-0.28x + 0.12y = 0$$
$$0.28x - 0.12y = 0$$
$$x + \quad y = 1$$

Solving, we find $x = 0.3$ and $y = 0.7$, and conclude that in the long run, there will be 30% 1-wage earners and 70% 2-wage earners.

21. a. If this trend continues, the percentage of homeowners in this city who will own single-family homes or condominiums two years from now will be given by $X_2 = TX_1$. Thus,

$$X_1 = TX_0 = \begin{bmatrix} .85 & .35 \\ .15 & .65 \end{bmatrix} \begin{bmatrix} .8 \\ .2 \end{bmatrix} = \begin{bmatrix} .75 \\ .25 \end{bmatrix}$$

$$X_2 = TX_1 = \begin{bmatrix} .85 & .35 \\ .25 & .65 \end{bmatrix} \begin{bmatrix} .75 \\ .25 \end{bmatrix} = \begin{bmatrix} .725 \\ .275 \end{bmatrix}$$

and we conclude that 72.5% will own single-family home and 27.5% will own condominiums at that time.

b. We solve the system

$$\begin{bmatrix} .85 & .35 \\ .15 & .65 \end{bmatrix} \begin{bmatrix} x \\ y \end{bmatrix} = \begin{bmatrix} x \\ y \end{bmatrix}$$

together with the equation $x + y = 1$. Thus,

$$-0.15x + 0.35y = 0$$
$$0.15x - 0.35y = 0$$
$$x + \quad y = 1$$

Solving, we find $x = 0.7$ and $y = 0.3$, and conclude that in the long run 70% will own single family homes and 30% will own condominiums.

23. a.

$$X_1 = TX_0 = \begin{bmatrix} .8 & .1 & .1 \\ .1 & .85 & .05 \\ .1 & .05 & .85 \end{bmatrix} \begin{bmatrix} .3 \\ .4 \\ .3 \end{bmatrix} = \begin{bmatrix} .31 \\ .385 \\ .305 \end{bmatrix}$$

$$X_2 = TX_1 = \begin{bmatrix} .8 & .1 & .1 \\ .1 & .85 & .05 \\ .1 & .05 & .85 \end{bmatrix} \begin{bmatrix} .31 \\ .385 \\ .305 \end{bmatrix} = \begin{bmatrix} .317 \\ .3735 \\ .3095 \end{bmatrix}$$

From our computations, we conclude that after two weeks 31.7% of the viewers will watch the *ABC* news, 37.35% will watch the *CBS* news, and 30.95% will watch the *NBC* news.

b. We solve the system

$$\begin{bmatrix} .8 & .1 & .1 \\ .1 & .85 & .05 \\ .1 & .05 & .85 \end{bmatrix} \begin{bmatrix} x \\ y \\ z \end{bmatrix} = \begin{bmatrix} x \\ y \\ z \end{bmatrix}$$

together with the equation $x + y + z = 1$, or equivalently, the system

$$-0.2x + 0.1y + 0.1z = 0$$
$$0.1x - 0.15y + 0.05z = 0$$
$$0.1x + 0.05y - 0.15z = 0$$
$$x + y + z = 1$$

Using the Gauss-Jordan elimination method, we find that the required steady-state

vector is $\begin{bmatrix} \frac{1}{3} \\ \frac{1}{3} \\ \frac{1}{3} \end{bmatrix}$, and conclude that each network will comand 33 1/3 % of the audience

in the long run.

25. We wish to solve

$$\begin{bmatrix} \frac{1}{2} & \frac{1}{4} & 0 \\ \frac{1}{2} & \frac{1}{2} & \frac{1}{2} \\ 0 & \frac{1}{4} & \frac{1}{2} \end{bmatrix} \begin{bmatrix} x \\ y \\ z \end{bmatrix} = \begin{bmatrix} x \\ y \\ z \end{bmatrix}$$

together with the equation $x + y + z = 1$, or, equivalently, the system of equations
$$\frac{1}{2}x + \frac{1}{4}y = x$$
$$\frac{1}{2}x + \frac{1}{2}y + \frac{1}{2}z = y$$
$$\frac{1}{4}y + \frac{1}{2}z = z$$
$$x + y + z = 1$$

Solving this system, we find that
$$x = \frac{1}{4}, \ y = \frac{1}{2}, \text{ and } z = \frac{1}{4}.$$

Thus, in the long run, 25% of the plants will have red flowers, 50% will have pink flowers and 25% will have white flowers.

27. False. All the entries of the limiting matrix must be positive as well.

29. Let T be a regular stochastic matrix and X the steady-state distribution vector that satisfies the equation $TX = T$ and assume that the sum of the elements of X are equal

to 1. Then $TX = X$ implies that $TX = T^2X$, or $X = T^2X, \dots$ So we have $X = T^n X$.

Next, let L be the steady-state distribution vector, then

$$L = \lim_{m \to \infty} X_m = \lim_{m \to \infty} T^m X_0 = T^m X_0$$

when m is large. Multiplying both sides by T, we obtain

$$TL = T^{m+1} X_0 \approx L.$$

Thus, L also satisfies $TL = L$ together with the condition that the sum of the elements in L be equal to 1. Since the matrix equation $TX = X$ has a unique solution, we conclude that $X = L$.

USING TECHNOLOGY EXERCISES 9.2, page 565

1. $X_5 = \begin{bmatrix} 0.2045 \\ 0.1319 \\ 0.2610 \\ 0.1868 \\ 0.2158 \end{bmatrix}$

EXERCISES 9.3, page 565

1. The given matrix is an absorbing stochastic matrix

$$T = \begin{matrix} & \begin{matrix} 1 & 2 \end{matrix} \\ \begin{matrix} 1 \\ 2 \end{matrix} & \begin{bmatrix} \frac{2}{5} & 0 \\ \frac{3}{5} & 1 \end{bmatrix} \end{matrix}$$

State 2 is an absorbing state. State 1 is nonabsorbing, but an object in this state has a probability of 3/5 of going to the absorbing state 2.

3. The given matrix is $\begin{matrix} & \begin{matrix} 1 & 2 & 3 \end{matrix} \\ \begin{matrix} 1 \\ 2 \\ 3 \end{matrix} & \begin{bmatrix} 1 & .5 & 0 \\ 0 & 0 & 1 \\ 0 & .5 & 0 \end{bmatrix} \end{matrix}$

States 1 and 3 are absorbing states. State 2 is not absorbing, but an object in this state has a probability of .5 of going to the absorbing state 1 and .5 of going to the absorbing state 3. Thus, the matrix is an absorbing matrix.

5. Yes. It is an absorbing stochastic matrix since it is possible to go from state 1 to the absorbing states 2 and 3.

7. The given matrix is

$$
\begin{array}{c c c c c}
 & 1 & 2 & 3 & 4 \\
\begin{array}{c} 1 \\ 2 \\ 3 \\ 4 \end{array} &
\left[\begin{array}{cccc}
1 & 0 & .3 & 0 \\
0 & 1 & .2 & 0 \\
0 & 0 & .1 & .5 \\
0 & 0 & .4 & .5
\end{array}\right]
\end{array}
$$

States 1 and 2 are absorbing states. States 3 and 4 are not. However, it is possible for an object to go from state 3 to state 1 with probability 0.3. furthermore, it is also possible for an object to go from the non-absorbing state 4 to an absorbing state. For example, via state 3 with a probability of 0.5. Therefore, the given matrix is an absorbing matrix.

9. The required matrix is
$$
\begin{array}{c c c}
 & 2 & 1 \\
\begin{array}{c} 2 \\ 1 \end{array} &
\left[\begin{array}{c|c}
1 & .4 \\
\hline
0 & .6
\end{array}\right]
\end{array}
$$
where $S = [.4]$ and $R = [.6]$

11.
$$
\begin{array}{c c c c}
 & 3 & 2 & 1 \\
\begin{array}{c} 3 \\ 2 \\ 1 \end{array} &
\left[\begin{array}{c|cc}
1 & .4 & .5 \\
\hline
0 & .4 & .5 \\
0 & .2 & 0
\end{array}\right]
\end{array}
$$
where $S = [.4 \quad .5]$ and $R = \begin{bmatrix} .4 & .5 \\ .2 & 0 \end{bmatrix}$

or
$$
\begin{array}{c c c c}
 & 3 & 1 & 2 \\
\begin{array}{c} 3 \\ 1 \\ 2 \end{array} &
\left[\begin{array}{c|cc}
1 & .5 & .4 \\
\hline
0 & 0 & .2 \\
0 & .5 & .4
\end{array}\right]
\end{array}
$$
where $S = [.5 \quad .4]$ and $R = \begin{bmatrix} 0 & .2 \\ .5 & .4 \end{bmatrix}$.

13.
$$
\left[\begin{array}{cc|cc}
1 & 0 & .2 & .4 \\
0 & 1 & .3 & 0 \\
\hline
0 & 0 & .3 & .2 \\
0 & 0 & .2 & .4
\end{array}\right]
, \quad
S = \begin{bmatrix} .2 & .4 \\ .3 & 0 \end{bmatrix}, \quad
R = \begin{bmatrix} .3 & .2 \\ .2 & .4 \end{bmatrix}
$$

$$\text{or} \quad \begin{bmatrix} 1 & 0 & \vdots & .4 & .2 \\ 0 & 1 & \vdots & 0 & .3 \\ \text{-} & \text{-} & \text{+} & \text{-} & \text{-} \\ 0 & 0 & \vdots & .4 & .2 \\ 0 & 0 & \vdots & .2 & .3 \end{bmatrix}, \quad S = \begin{bmatrix} .4 & .2 \\ 0 & .3 \end{bmatrix}, \quad R = \begin{bmatrix} .4 & .2 \\ .2 & .3 \end{bmatrix}$$

and so forth.

15. Rewriting the matrix so that the absorbing states appear first, we have

$$\begin{array}{cc} & \begin{array}{cc} 2 & 1 \end{array} \\ \begin{array}{c} 2 \\ 1 \end{array} & \begin{bmatrix} 1 & \vdots & .45 \\ \text{-} & \text{+} & \text{-} \\ 0 & \vdots & .55 \end{bmatrix} \end{array} \quad \text{where } S = [.45] \text{ and } R = [.55]. \text{ Then}$$

$$(I - R) = [.45] \text{ and } S(I - R)^{-1} = [.45]\left[\dfrac{1}{.45}\right] = 1$$

Therefore the steady-state matrix is $\begin{array}{cc} & \begin{array}{cc} 2 & 1 \end{array} \\ \begin{array}{c} 2 \\ 1 \end{array} & \begin{bmatrix} 1 & \vdots & 1 \\ \text{-} & \text{+} & \text{-} \\ 0 & \vdots & 0 \end{bmatrix} \end{array}$.

17. Here we have $\begin{bmatrix} 1 & \vdots & .2 & .3 \\ \text{-} & \text{+} & \text{-} & \text{-} \\ 0 & \vdots & .4 & .2 \\ 0 & \vdots & .4 & .5 \end{bmatrix}$ where $S = [.2 \quad .3]$ and $R = \begin{bmatrix} .4 & .4 \\ .2 & .5 \end{bmatrix}$

Next, $I - R = \begin{bmatrix} 1 & 0 \\ 0 & 1 \end{bmatrix} - \begin{bmatrix} .4 & .2 \\ .4 & .5 \end{bmatrix} = \begin{bmatrix} .6 & -.2 \\ -.4 & .5 \end{bmatrix}$

Using the formula for finding the inverse of a 2×2 matrix, we have

$$(I - R)^{-1} = \begin{bmatrix} 2.27 & .91 \\ 1.8 & 2.73 \end{bmatrix}$$

Then $S(I - R)^{-1} = [.2 \quad .3]\begin{bmatrix} 2.27 & .91 \\ 1.8 & 2.73 \end{bmatrix} = [.994 \quad 1] \approx [1 \quad 1]$

We conclude that the steady-state matrix is $\begin{bmatrix} 1 & \vdots & 1 & 1 \\ \text{-} & \text{+} & \text{-} & \text{-} \\ 0 & \vdots & 0 & 0 \\ 0 & \vdots & 0 & 0 \end{bmatrix}$.

19. Upon rewriting the given matrix so that the absorbing states appear first, we have

$$
\begin{array}{c}
\begin{array}{cccc} 2 & 4 & 1 & 3 \end{array} \\
\begin{array}{c} 2 \\ 4 \\ 1 \\ 3 \end{array}
\left[
\begin{array}{cc:cc}
1 & 0 & \frac{1}{2} & 0 \\
0 & 1 & 0 & 0 \\
\hdashline
0 & 0 & \frac{1}{2} & \frac{1}{3} \\
0 & 0 & 0 & \frac{2}{3}
\end{array}
\right]
\end{array}
$$

where $S = \begin{bmatrix} \frac{1}{2} & 0 \\ 0 & 0 \end{bmatrix}$ and $R = \begin{bmatrix} \frac{1}{2} & \frac{1}{3} \\ 0 & \frac{2}{3} \end{bmatrix}$. Next, we compute

$$
I - R = \begin{bmatrix} 1 & 0 \\ 0 & 1 \end{bmatrix} - \begin{bmatrix} \frac{1}{2} & \frac{1}{3} \\ 0 & \frac{2}{3} \end{bmatrix} = \begin{bmatrix} \frac{1}{2} & -\frac{1}{3} \\ 0 & \frac{1}{3} \end{bmatrix}.
$$

Using the formula for finding the inverse of a 2×2 matrix, we have

$$
(I - R)^{-1} = \begin{bmatrix} 2 & 2 \\ 0 & 3 \end{bmatrix} \text{ and so } S(I - R)^{-1} = \begin{bmatrix} 2 & 2 \\ 0 & 3 \end{bmatrix} \begin{bmatrix} \frac{1}{2} & 0 \\ 0 & 0 \end{bmatrix} = \begin{bmatrix} 1 & 1 \\ 0 & 0 \end{bmatrix}.
$$

Therefore, the steady-state matrix is
$\begin{bmatrix}
1 & 0 & 1 & 1 \\
0 & 1 & 0 & 0 \\
\hdashline
0 & 0 & 0 & 0 \\
0 & 0 & 0 & 0
\end{bmatrix}$.

21. Here
$\begin{bmatrix}
1 & 0 & \frac{1}{4} & \frac{1}{3} \\
0 & 1 & \frac{1}{4} & \frac{1}{3} \\
\hdashline
0 & 0 & \frac{1}{2} & 0 \\
0 & 0 & 0 & \frac{1}{3}
\end{bmatrix}$,
$S = \begin{bmatrix} \frac{1}{4} & \frac{1}{3} \\ \frac{1}{4} & \frac{1}{3} \end{bmatrix}$,
$R = \begin{bmatrix} \frac{1}{2} & 0 \\ 0 & \frac{1}{3} \end{bmatrix}$

and $I - R = \begin{bmatrix} 1 & 0 \\ 0 & 1 \end{bmatrix} - \begin{bmatrix} \frac{1}{2} & 0 \\ 0 & \frac{1}{3} \end{bmatrix} = \begin{bmatrix} \frac{1}{2} & 0 \\ 0 & \frac{2}{3} \end{bmatrix}.$

Using the formula for finding the inverse of a 2×2 matrix, we find

$$
(I - R)^{-1} = \begin{bmatrix} 2 & 0 \\ 0 & \frac{3}{2} \end{bmatrix} \text{ and } S(I - R)^{-1} = \begin{bmatrix} \frac{1}{4} & \frac{1}{3} \\ \frac{1}{4} & \frac{1}{3} \end{bmatrix} \begin{bmatrix} 2 & 0 \\ 0 & \frac{3}{2} \end{bmatrix} = \begin{bmatrix} \frac{1}{2} & \frac{1}{2} \\ \frac{1}{2} & \frac{1}{2} \end{bmatrix}.
$$

The steady-state matrix is given by $\begin{bmatrix} 1 & 0 & \vdots & \frac{1}{2} & \frac{1}{2} \\ 0 & 1 & \vdots & \frac{1}{2} & \frac{1}{2} \\ \cdots & \cdots & \vdots & \cdots & \cdots \\ 0 & 0 & \vdots & 0 & 0 \\ 0 & 0 & \vdots & 0 & 0 \end{bmatrix}$.

23. The absorbing states already appear first in the matrix, so it need not be rewritten. Next, $(I - R) = \begin{bmatrix} .8 & -.2 \\ -.2 & .6 \end{bmatrix}$ and $(I - R)^{-1} = \begin{bmatrix} \frac{15}{11} & \frac{5}{11} \\ \frac{5}{11} & \frac{20}{11} \end{bmatrix}$ so that

$$S(I - R)^{-1} = \begin{bmatrix} \frac{2}{10} & \frac{1}{10} \\ \frac{1}{10} & \frac{2}{10} \\ \frac{3}{10} & \frac{1}{10} \end{bmatrix} \begin{bmatrix} \frac{15}{11} & \frac{5}{11} \\ \frac{5}{11} & \frac{20}{11} \end{bmatrix} = \begin{bmatrix} \frac{7}{22} & \frac{3}{11} \\ \frac{5}{22} & \frac{9}{22} \\ \frac{5}{11} & \frac{7}{22} \end{bmatrix}$$

Therefore, the steady-state matrix is given by

$$\begin{bmatrix} 1 & 0 & 0 & \vdots & \frac{7}{22} & \frac{3}{11} \\ 0 & 1 & 0 & \vdots & \frac{5}{22} & \frac{9}{22} \\ 0 & 0 & 1 & \vdots & \frac{5}{11} & \frac{7}{22} \\ \cdots & \cdots & \cdots & \vdots & \cdots & \cdots \\ 0 & 0 & 0 & \vdots & 0 & 0 \\ 0 & 0 & 0 & \vdots & 0 & 0 \end{bmatrix} .$$

25. State 2 is absorbing. State 1 is not absorbing, but it is possible for an object to go from state 1 to state 2 with probability 0.8. Therefore, the matrix is absorbing. Rewriting, we obtain

$$\begin{array}{c} \quad 2 \quad 1 \\ \begin{array}{c} 1 \\ 2 \end{array} \begin{bmatrix} 1 & \vdots & .2 \\ \cdots & \vdots & \cdots \\ 0 & \vdots & .8 \end{bmatrix} \end{array}$$ where $S = [0.2]$ and $R = [0.8]$.

b. We compute $I - R = [1] - [0.2] = [0.8]$. So $(I - R)^{-1} = [1.25]$. Therefore,

$$S(I - R)^{-1} = [.8][1.25] = [1]$$ and the steady state matrix is $\begin{bmatrix} 1 & \vdots & 1 \\ \cdots & \vdots & \cdots \\ 0 & \vdots & 0 \end{bmatrix}$

this result tells us that in the long run only unleaded gas will be used.

27. Here

$$
\begin{array}{c}
\ \ \ \$0\ \ \$4\ \$1\ \$2\ \$3 \\
\begin{array}{c}\$0\\\$4\\\$1\\\$2\\\$3\end{array}
\left[\begin{array}{ccc:cc}
1 & 0 & \frac{1}{2} & 0 & 0 \\
0 & 1 & 0 & 0 & \frac{1}{2} \\
\hdashline
0 & 0 & 0 & \frac{1}{2} & 0 \\
0 & 0 & \frac{1}{2} & 0 & \frac{1}{2} \\
0 & 0 & 0 & \frac{1}{2} & 0
\end{array}\right]
\end{array}
\quad \text{where } S = \left[\begin{array}{ccc} \frac{1}{2} & 0 & 0 \\ 0 & 0 & \frac{1}{2} \end{array}\right] \quad \text{and } R = \left[\begin{array}{ccc} 0 & \frac{1}{2} & 0 \\ \frac{1}{2} & 0 & \frac{1}{2} \\ 0 & \frac{1}{2} & 0 \end{array}\right].
$$

Next, $I - R = \left[\begin{array}{ccc} 1 & -\frac{1}{2} & 0 \\ -\frac{1}{2} & 1 & -\frac{1}{2} \\ 0 & -\frac{1}{2} & 1 \end{array}\right]$ and $(I - R)^{-1} = \left[\begin{array}{ccc} \frac{3}{2} & 1 & \frac{3}{2} \\ 1 & 2 & 1 \\ \frac{3}{2} & 1 & \frac{3}{2} \end{array}\right]$ and

$$
S(I - R)^{-1} = \left[\begin{array}{ccc} \frac{3}{4} & \frac{1}{2} & \frac{1}{4} \\ \frac{1}{4} & \frac{1}{2} & \frac{3}{4} \end{array}\right].
$$

Therefore, the steady-state matrix is given by
$\left[\begin{array}{cc:ccc}
1 & 0 & \frac{3}{4} & \frac{1}{2} & \frac{1}{4} \\
0 & 1 & \frac{1}{4} & \frac{1}{2} & \frac{3}{4} \\
\hdashline
0 & 0 & 0 & 0 & 0 \\
0 & 0 & 0 & 0 & 0 \\
0 & 0 & 0 & 0 & 0
\end{array}\right]$

We conclude that if Diane started out with $1, the probability that she would leave the game a winner is 1/4. Similarly, if she started out with $2, the probability that she would leave the game a winner is 1/2, and if she started out with $3, the probability that she would leave as a winner is 3/4.

29.

$$
\left[\begin{array}{cc:cc}
1 & 0 & .25 & .1 \\
0 & 1 & 0 & .9 \\
\hdashline
0 & 0 & 0 & 0 \\
0 & 0 & .75 & 0
\end{array}\right]
$$

b. $I - R = \begin{bmatrix} 1 & 0 \\ -.75 & 1 \end{bmatrix}$ and $(I - R)^{-1} = \begin{bmatrix} 1 & 0 \\ .75 & 1 \end{bmatrix}$ and

$S(I - R)^{-1} = \begin{bmatrix} .25 & .1 \\ 0 & .9 \end{bmatrix} \begin{bmatrix} 1 & 0 \\ .75 & 1 \end{bmatrix} = \begin{bmatrix} .325 & .1 \\ .675 & .9 \end{bmatrix}$.

Therefore, the steady-state matrix is

$$\begin{bmatrix} 1 & 0 & .325 & .1 \\ 0 & 1 & .675 & .9 \\ 0 & 0 & 0 & 0 \\ 0 & 0 & 0 & 0 \end{bmatrix}$$

c. From the steady-state matrix, we see that the probability that a beginning student enrolled in the program will compete the course successfullly is 0.675.

31. False. It must be possible to go from any nonabsorbing state to an absorbing state in one or more stages.

33. The transition matrix is
$$\begin{matrix} & aa\ Aa\ AA \\ \begin{matrix} aa \\ Aa \\ AA \end{matrix} & \begin{bmatrix} 1 & \frac{1}{2} & 0 \\ 0 & \frac{1}{2} & 1 \\ 0 & 0 & 0 \end{bmatrix} \end{matrix}.$$
Since the entries in T are exactly the

same as those in Example 4, the steady-state matrix is

$$\begin{matrix} & aa\ Aa\ AA \\ \begin{matrix} aa \\ Aa \\ AA \end{matrix} & \begin{bmatrix} 1 & 1 & 1 \\ 0 & 0 & 0 \\ 0 & 0 & 0 \end{bmatrix} \end{matrix}.$$

Interpreting the steady-state matrix, we see that in the long run all the flowers produced by the plants will be white.

EXERCISES 9.4, page 577

1. We first determine the minimum of each row and the maxima of each column of the payoff matrix. Next, we find the larger of the row minima and the smaller of the column maxima as shown in the following matrix.

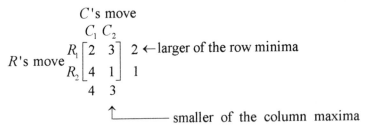

C's move

$$
\begin{array}{c}
\quad\quad C_1 \; C_2 \\
R\text{'s move} \begin{array}{c} R_1 \\ R_2 \end{array} \begin{bmatrix} 2 & 3 \\ 4 & 1 \end{bmatrix} \begin{array}{c} 2 \\ 1 \end{array} \leftarrow \text{larger of the row minima} \\
\quad\quad\; 4 \;\; 3
\end{array}
$$

↑———— smaller of the column maxima

From the above results, we conclude that the row player's maximum strategy is to play row 1, whereas the column player's minimax strategy is to play column 2.

3. We first obtain the following matrix, where the larger of the row minima and the smallest of the column maxima are displayed.

$$
\begin{array}{c}
\quad\quad C_1 \;\; C_2 \;\; C_3 \\
\begin{array}{c} R_1 \\ R_2 \end{array} \begin{bmatrix} 1 & 3 & 2 \\ 0 & -1 & 4 \end{bmatrix} \begin{array}{c} 1 \\ -1 \end{array} \leftarrow \text{larger of the row minima} \\
\quad\quad\;\; 1 \;\;\;\; 3 \;\;\;\; 4
\end{array}
$$

↑—— smallest of the column maxima

We conclude that the row player's maximum strategy is to play row 1, whereas the column player's minimax strategy is to play column 1.

5. From the following payoff matrix where the largest of the row minima and the smallest of the column maxima are displayed

$$
\begin{bmatrix} 3 & 2 & 1 \\ 1 & -2 & 3 \\ 6 & 4 & 1 \end{bmatrix}
\begin{array}{c} 1 \;\leftarrow \\ -2 \\ 1 \;\leftarrow \end{array}
\quad \text{largest of the row minima}
$$

$$
\quad\quad 6 \;\;\; 4 \;\;\; 3
$$

↑—— smallest of the column maxima

we conclude that the row player's maximin strategy is to play either row 1 or row 3, whereas the column player's minimax strategy is to play column 3.

7. From the following payoff matrix, where the largest of the row minima and the

smallest of the column maxima are displayed,

$$\begin{bmatrix} 4 & 2 & 1 \\ 1 & 0 & -1 \\ 2 & 1 & 3 \end{bmatrix} \begin{matrix} 1\leftarrow \\ -1 \\ 1\leftarrow \end{matrix}$$ — largest of the row minima

$$\begin{matrix} 4 & 2 & 3 \end{matrix}$$

↑ smallest of the column maxima

we conclude that the row player's maximin strategy is to play either row 1 or row 3, whereas the column player's minimax strategy is to play column 2.

9. From

$$\begin{bmatrix} ② & 3 \\ 1 & -4 \end{bmatrix} \begin{matrix} 2\leftarrow \text{larger of the row minima} \\ -4 \end{matrix}$$

$$\begin{matrix} 2 & 3 \end{matrix}$$

↑ smallest of the column maxima

we see that the game is strictly determined, and
a. the saddle point is 2.
b. the optimum strategy for the row player is to play row 1, whereas the optimum srategy for the column player is to play column 1.
c. the value of the game is 2.
d. the game favors the row player.

11. From

$$\begin{bmatrix} ① & 3 & 2 \\ -1 & 4 & -6 \end{bmatrix} \begin{matrix} 1\leftarrow \text{larger of the row minima} \\ -6 \end{matrix}$$

$$\begin{matrix} 1 & 4 & 2 \end{matrix}$$

↑ smallest of the column maxima

we see that the game is strictly determined, and
a. the saddle point is 1.
b. the optimum strategy for the row player is to play row 1, whereas, the optimum strategy for the column player is to play column 1.
c. the value of the game is 1.
d. the game favors the row player.

13. From

$$\begin{bmatrix} ① & 3 & 4 & 2 \\ 0 & 2 & 6 & -4 \\ -1 & -3 & -2 & -1 \end{bmatrix} \begin{matrix} 1 \leftarrow \text{largest of the row minima} \\ -4 \\ -3 \end{matrix}$$

$$\begin{matrix} 1 & 3 & 6 & 2 \\ \uparrow \underline{\hspace{5cm}} & & & \text{smallest of the column maxima} \end{matrix}$$

we conclude that the game is strictly determined, and
a. the saddle point is 1.
b. the optimum strategy for the row player is to play row 1, while the optimum strategy for the column player is to play column 1.
c. the value of the game is 1.
d. the game favors the row player.

15. From

$$\begin{bmatrix} 1 & 2 \\ 0 & 3 \\ -1 & 2 \\ 2 & -2 \end{bmatrix} \begin{matrix} 1 \leftarrow \text{larger of the row minima} \\ 0 \\ -1 \\ -2 \end{matrix}$$

$$\begin{matrix} 2 & 3 \\ \uparrow \underline{\hspace{3cm}} & \text{smaller of the column maxima} \end{matrix}$$

we see that the game is not strictly determined and consequently has no saddle point.

17. From

$$\begin{bmatrix} 1 & -1 & 3 & 2 \\ 1 & 0 & 2 & 2 \\ -2 & 2 & 3 & -1 \end{bmatrix} \begin{matrix} -1 \\ 0 \leftarrow \text{largest of the row minima} \\ -2 \end{matrix}$$

$$\begin{matrix} 1 & 2 & 3 & 2 \\ \uparrow \underline{\hspace{4cm}} & & & \text{smallest of the column maxima} \end{matrix}$$

we conclude that the game is not strictly determined since there is no entry that is simultaneously the largest of the row minima and the smallest of the column maxima.

19. a.

$$\begin{array}{c c c c}
 & 1 & 2 & 3 \\
1 & \begin{bmatrix} 2 & -3 & 4 \\ 2 & -3 & 4 & -5 \\ 3 & 4 & -5 & 6 \end{bmatrix}
\end{array}$$

b. From

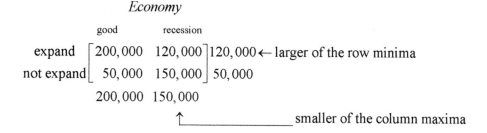

$$\begin{bmatrix} 2 & -3 & 4 \\ -3 & 4 & -5 \\ 4 & -5 & 6 \end{bmatrix} \begin{matrix} -3 \leftarrow \text{largest of the row minima} \\ -5 \\ -5 \end{matrix}$$

$$\qquad 4 \quad 4 \quad 6$$

$$\qquad \uparrow \quad \uparrow \underline{\qquad\qquad} \text{smaller of the column maxima}$$

we conclude that the maximin strategy for Robin is to play row 1 (extend 1 finger), whereas the minimax strategy for Cathy is to play column 1 or column 2 (extend 1 or 2 fingers).

21. a. The following is the payoff matrix for the game.

Economy

$$\begin{array}{c c c}
 & \text{good} & \text{recession} \\
\text{expand} & \begin{bmatrix} 200,000 & 120,000 \\ \text{not expand} & 50,000 & 150,000 \end{bmatrix} & \begin{matrix} 120,000 \leftarrow \text{larger of the row minima} \\ 50,000 \end{matrix}
\end{array}$$

$$\qquad 200,000 \quad 150,000$$

$$\qquad \uparrow\underline{\qquad\qquad} \text{smaller of the column maxima}$$

b. The row player's (management) minimax strategy is to play row 1, that is, to expand its line of conventional speakers.

23. a. The following is the payoff matrix for this game.

$$
\begin{array}{cc}
 & Charley's \\
 & \begin{array}{ccc} R & H & L \end{array} \\
\begin{array}{c} R \\ Roland's\ H \\ L \end{array} &
\left[\begin{array}{ccc}
3 & -1 & -3 \\
2 & 0 & -2 \\
5 & 2 & ① \\
\end{array}\right]
\begin{array}{c} -3 \\ -2 \\ 1 \leftarrow \text{largest of the row minima} \end{array} \\
 & \begin{array}{ccc} 5 & 2 & 1 \end{array}
\end{array}
$$

$\uparrow$ _____ smallest of the column maxima

b. From the payoff matrix, we see that the game is strictly determined.

c. If neither party is willing to lower their price, the payoff matrix would be

$$
\begin{array}{cc}
 & Charley's \\
 & \begin{array}{cc} R & C \end{array} \\
Roland's\ \begin{array}{c} R \\ C \end{array} &
\left[\begin{array}{cc}
3 & -1 \\
2 & ⓪ \\
\end{array}\right]
\begin{array}{c} -1 \\ 0 \leftarrow \text{larger of the row minima} \end{array} \\
 & \begin{array}{cc} 3 & 0 \end{array}
\end{array}
$$

$\uparrow$ _____ smaller of the column maxima

and we see that the game is strictly determined, so that the optimal strategy for each barber is to charge his current price for a haircut.

25. True. This follows from the defintion.

EXERCISES 9.5, page 589

1. We compute

$$
E = PAQ = \begin{bmatrix} \frac{1}{2} & \frac{1}{2} \end{bmatrix} \begin{bmatrix} 3 & 1 \\ -4 & 2 \end{bmatrix} \begin{bmatrix} \frac{3}{5} \\ \frac{2}{5} \end{bmatrix}
$$

$$
= \begin{bmatrix} -\frac{1}{2} & \frac{3}{2} \end{bmatrix} \begin{bmatrix} \frac{3}{5} \\ \frac{2}{5} \end{bmatrix} = -\frac{3}{10} + \frac{6}{10} = \frac{3}{10}.
$$

Thus, in the long run, the row player may be expected to win 0.3 units in each play of the game.

3. We compute

$$E = PAQ = \begin{bmatrix} \frac{1}{3} & \frac{2}{3} \end{bmatrix} \begin{bmatrix} -4 & 3 \\ 2 & 1 \end{bmatrix} \begin{bmatrix} \frac{3}{4} \\ \frac{1}{4} \end{bmatrix}$$

$$= \begin{bmatrix} 0 & \frac{5}{3} \end{bmatrix} \begin{bmatrix} \frac{3}{4} \\ \frac{1}{4} \end{bmatrix} = \frac{5}{12}.$$

Thus, in the long run, the row player may be expected to win 0.4167 units in each play of the game.

5. We compute

$$E = PAQ = \begin{bmatrix} .02 & .06 & .02 \end{bmatrix} \begin{bmatrix} 2 & 0 & -2 \\ 1 & -1 & 3 \\ 2 & 1 & -4 \end{bmatrix} \begin{bmatrix} 0.2 \\ 0.6 \\ 0.2 \end{bmatrix}$$

$$= \begin{bmatrix} 1.4 & -0.4 & 0.6 \end{bmatrix} \begin{bmatrix} 0.2 \\ 0.6 \\ 0.2 \end{bmatrix} = 0.16.$$

7. a. $\quad E = PAQ = \begin{bmatrix} 1 & 0 \end{bmatrix} \begin{bmatrix} 1 & -2 \\ -2 & 3 \end{bmatrix} \begin{bmatrix} 1 \\ 0 \end{bmatrix} = \begin{bmatrix} 1 & -2 \end{bmatrix} \begin{bmatrix} 1 \\ 0 \end{bmatrix} = 1$

   b. $\quad E = PAQ = \begin{bmatrix} 0 & 1 \end{bmatrix} \begin{bmatrix} 1 & -2 \\ -2 & 3 \end{bmatrix} \begin{bmatrix} 1 \\ 0 \end{bmatrix} = \begin{bmatrix} -2 & 3 \end{bmatrix} \begin{bmatrix} 1 \\ 0 \end{bmatrix} = -2.$

   c.

   $$E = PAQ = \begin{bmatrix} \frac{1}{2} & \frac{1}{2} \end{bmatrix} \begin{bmatrix} 1 & -2 \\ -2 & 3 \end{bmatrix} \begin{bmatrix} \frac{1}{2} \\ \frac{1}{2} \end{bmatrix}$$

   $$= \begin{bmatrix} -\frac{1}{2} & \frac{1}{2} \end{bmatrix} \begin{bmatrix} \frac{1}{2} \\ \frac{1}{2} \end{bmatrix} = 0.$$

   d.

   $$E = PAQ = \begin{bmatrix} 0.5 & 0.5 \end{bmatrix} \begin{bmatrix} 1 & -2 \\ -2 & 3 \end{bmatrix} \begin{bmatrix} 0.8 \\ 0.2 \end{bmatrix}$$

   $$= \begin{bmatrix} -0.5 & 0.5 \end{bmatrix} \begin{bmatrix} 0.8 \\ 0.2 \end{bmatrix} = -0.3.$$

   (a) is the most advantageous.

9. a. From the payoff matrix

$$\begin{bmatrix} -3 & 3 & 2 \\ -3 & 1 & 1 \\ 1 & -2 & 1 \\ & & \end{bmatrix} \begin{matrix} -3 \\ -3 \\ -2 \\ \end{matrix}$$ _____ largest of the row minima

$$\begin{matrix} 1 & 3 & 2 \\ \uparrow & & \end{matrix}$$ _____ smallest of the column maxima

we see that the expected payoff to a row player using the minimax strategy is 1.

b. The expected payoff is given by

$$E = PAQ = \begin{bmatrix} .25 & .25 & .5 \end{bmatrix} \begin{bmatrix} -3 & 3 & 2 \\ -3 & 1 & 1 \\ 1 & -2 & 1 \end{bmatrix} \begin{bmatrix} .6 \\ .2 \\ .2 \end{bmatrix} = -0.35$$

c. The minimax strategy (part (a)) is the better strategy for the row player.

11. The game under consideration has no saddle point and is accordingly nonstrictly determined. Using the formulas for determing the optimal mixed strategies for a 2×2 game with $a = 4$, $b = 1$, $c = 2$, and $d = 3$, we find that

$$p_1 = \frac{d-c}{a+d-b-c} = \frac{3-2}{4+3-1-2} = \frac{1}{4}$$

$$p_2 = 1 - p_1 = 1 - \frac{1}{4} = \frac{3}{4},$$

so that the row player's optimal mixed strategy is given by

$$P = \begin{bmatrix} \frac{1}{4} & \frac{3}{4} \end{bmatrix}.$$

Next, we compute

$$q_1 = \frac{d-b}{a+d-b-c} = \frac{3-1}{4+3-1-2} = \frac{2}{4} = \frac{1}{2}.$$

$$q_2 = 1 - q_1 = 1 - \frac{1}{2} = \frac{1}{2}.$$

Thus, the optimal strategy for the column player is given by

$$Q = \begin{bmatrix} \frac{1}{2} \\ \frac{1}{2} \end{bmatrix}.$$

To determine whether the game favors one player over the other, we compute the expected value of the game which is given by

$$E = \frac{ad - bc}{a+d-b-c} = \frac{(4)(3)-(1)(2)}{4+3-1-2} = \frac{10}{4} = \frac{5}{2},$$

or 5/2 units for each play of the game. These results imply that the game favors the row player.

13. Since the game is not strictly determined, we use the formulas for determing the optimal mixed strategies for a 2×2 game. We obtain

$$p_1 = \frac{d-c}{a+d-b-c} = \frac{-3-1}{-1-3-2-1} = \frac{4}{7}$$

$$p_2 = 1 - p_1 = 1 - \frac{4}{7} = \frac{3}{7}.$$

Thus, the optimal mixed strategy for the row player is given by
$$P = \begin{bmatrix} \frac{4}{7} & \frac{3}{7} \end{bmatrix}.$$

To find the optimal mixed strategy for the column player, we compute

$$q_1 = \frac{d-b}{a+d-b-c} = \frac{-3-2}{-1-3-2-1} = \frac{5}{7}$$

$$q_2 = 1 - q_1 = 1 - \frac{5}{7} = \frac{2}{7}.$$

Hence, $Q = \begin{bmatrix} \frac{5}{7} \\ \frac{2}{7} \end{bmatrix}.$

The expected value of the game is given by
$$E = \frac{ad - bc}{a+d-b-c} = \frac{(-1)(-3)-(2)(1)}{-1-3-2-1} = -\frac{1}{7}.$$

Since the value of the game is negative, we conclude that the game favors the column player.

15. Since the game is not strictly determined, we use the formulas for determining the optimal mixed strategies for a 2×2 game. We obtain

$$P = \begin{bmatrix} \frac{1}{2} & \frac{1}{2} \end{bmatrix} \quad \text{and} \quad Q = \begin{bmatrix} \frac{1}{4} \\ \frac{3}{4} \end{bmatrix}$$

and $E = PAQ = -5$. We conclude that the game favors the column player.

17. a. Since the game is not strictly determined, we employ the formulas for determining the optimal mixed strategies for a 2×2 game. We find that

$$P_1 = \frac{d-c}{a+d-b-c} = \frac{1-(-2)}{4+1-(-2)-(-2)} = \frac{3}{9} = \frac{1}{3}$$

$$P_2 = 1 - P_2 = 1 - \frac{1}{3} = \frac{2}{3},$$

so that Richie's optimal mixed strategy is given by

$$P = \begin{bmatrix} \frac{1}{3} & \frac{2}{3} \end{bmatrix}.$$

Next, we compute

$$q_1 = \frac{d-b}{a+d-b-c} = \frac{1-(-2)}{4+1-(-2)-(-2)} = \frac{3}{9} = \frac{1}{3}$$

$$q_2 = 1 - q_1 = 1 - \frac{1}{3} = \frac{2}{3}.$$

Thus, Chuck's optimal strategy is given by

$$Q = \begin{bmatrix} \frac{1}{3} \\ \frac{2}{3} \end{bmatrix}.$$

b. The expected value of the game is given by

$$E = \frac{ad - bc}{a+d-b-c} = \frac{(4)(1)-(-2)(-2)}{4+1-(-2)-(-2)} = 0$$

and conclude that in the long run the game will end in a draw.

19. a. The required payoff matrix for this game is given by

	Expanding economy	Economic recession
Hotel stock	25	−5
Brewery stock	10	15

Since the game is not strictly determined, we use the formulas for finding the optimal mixed strategies for a 2×2 nonstrictly determined game. Then

$$P_1 = \frac{d-c}{a+d-b-c} = \frac{15-10}{25+15+5-10} = \frac{5}{35} = \frac{1}{7}$$

$$P_2 = 1 - P_2 = 1 - \frac{1}{7} = \frac{6}{7},$$

so that the Maxwell's optimal mixed strategy is

$$P = \begin{bmatrix} \frac{1}{7} & \frac{6}{7} \end{bmatrix}$$

Thus, the Maxwells should invest $(1/7)(\$40,000) = \$5714$ in hotel stocks and $(6/7)(\$40,000) = \$34,286$ in brewery stocks.

b. The profit that the Maxwell's can expect to make is given by
$$E = \frac{ad - bc}{a + d - b - c} = \frac{(25)(15) - (-5)(10)}{35} = \frac{425}{35} = 12.148.$$
We conclude that the Maxwell's will realize a profit of
$$(0.12148)(\$40,000) = \$4857$$
by employing their optimal mixed strategy.

21. a. The required payoff matrix for this game is given by

$$
\begin{array}{cc}
& C \\
& \begin{array}{cc} N & F \end{array} \\
R \; \begin{array}{c} N \\ F \end{array} & \left[\begin{array}{cc} .48 & .65 \\ .50 & .45 \end{array} \right] \begin{array}{c} .48 \\ .45 \end{array} \\
& \begin{array}{cc} .50 & .65 \end{array}
\end{array}
\qquad
\begin{array}{l}
N = \text{local newspaper} \\
F = \text{flyer}
\end{array}
$$

Since there is no saddle point, we conclude that the game is not strictly determined.
b. Employing the formulas for finding the optimal mixed strategies for a 2×2 nonstrictly determined game, we find that

$$p_1 = \frac{d - c}{a + d - b - c} = \frac{.45 - .50}{.48 + .45 - .65 - .50} = .227$$
$$p_2 = 1 - p_1 = 1 - .227 = .773.$$
$$q_1 = \frac{d - b}{a + d - b - c} = \frac{.45 - .65}{.48 + .45 - .65 - .50} = .909$$
$$q_2 = 1 - q_1 = 1 - .909 = .091.$$

We conclude that Dr. Russell's strategy is given by
$$P = \begin{bmatrix} .227 & .773 \end{bmatrix}$$
and Dr. Carlton's strategy is given by
$$Q = \begin{bmatrix} .909 \\ .091 \end{bmatrix}.$$

Also, Dr. Russell should place 22.7% of his advertisements in the local newspaper and 77.3% in fliers; whereas, Dr. Carlton should place 90.9% of his advertisements in the local newspaper and 9.1% of his advertisements in fliers.

23. The optimal strategies for the row and column players are

$$P = \begin{bmatrix} p_1 & p_2 \end{bmatrix}$$

$$\text{where} \quad p_1 = \frac{d-c}{a+d-b-c} \quad \text{and}$$

$$p_2 = 1 - p_1 = 1 - \frac{d-c}{a+d-b-c} = \frac{a-b}{a+d-b-c}$$

$$\text{and} \quad Q = \begin{bmatrix} q_1 \\ q_2 \end{bmatrix}$$

$$\text{where} \quad q_1 = \frac{d-b}{a+d-b-c}$$

$$\text{and} \quad q_2 = 1 - q_1 = 1 - \frac{d-b}{a+d-b-c} = \frac{a-c}{a+d-b-c}.$$

Therefore, the expected value of the game is

$$E = PAQ = \begin{bmatrix} p_1 & p_2 \end{bmatrix} \begin{bmatrix} a & b \\ c & d \end{bmatrix} \begin{bmatrix} q_1 \\ q_2 \end{bmatrix}$$

$$= \begin{bmatrix} ap_1 + cp_2 & bp_1 + dp_2 \end{bmatrix} \begin{bmatrix} q_1 \\ q_2 \end{bmatrix}$$

$$= (ap_1 + cp_2)q_1 + (bp_1 + dp_2)q_2$$

$$= \left[\left[\frac{a(d-c)}{a+d-b-c} \right] + \left[\frac{c(a-b)}{a+d-b-c} \right] \right] \left[\frac{d-b}{a+d-b-c} \right]$$

$$+ \left[\left[\frac{b(d-c)}{a+d-b-c} \right] + \left[\frac{d(a-b)}{a+d-b-c} \right] \right] \left[\frac{(a-c)}{a+d-b-c} \right]$$

$$= \frac{(ad-bc)(d-b)}{(a+d-b-c)^2} + \frac{(ad-bc)(a-c)}{(a+d-b-c)^2}$$

$$= \frac{(ad-bc)(a+d-b-c)}{(a+d-b-c)^2} = \frac{ad-bc}{a+d-b-c},$$

as was to be shown.

CHAPTER 9, REVIEW EXERCISES, page 595

1. Since the entries $a_{12} = -2$ and $a_{22} = -8$ are negative, the given matrix is not stochastic and is hence not a regular stochastic matrix.

3. $T^2 = \begin{bmatrix} \frac{1}{2} & 0 & \frac{1}{3} \\ 0 & 0 & \frac{1}{3} \\ \frac{1}{2} & 1 & \frac{1}{3} \end{bmatrix} \begin{bmatrix} \frac{1}{2} & 0 & \frac{1}{3} \\ 0 & 0 & \frac{1}{3} \\ \frac{1}{2} & 1 & \frac{1}{3} \end{bmatrix} = \begin{bmatrix} \frac{5}{12} & \frac{1}{3} & \frac{5}{18} \\ \frac{1}{6} & \frac{1}{3} & \frac{1}{9} \\ \frac{5}{12} & \frac{1}{3} & \frac{11}{18} \end{bmatrix}$

and so the matrix is regular.

5. $X_1 = \begin{bmatrix} 0 & \frac{1}{4} & \frac{3}{5} \\ \frac{2}{5} & \frac{1}{2} & \frac{1}{5} \\ \frac{3}{5} & \frac{1}{4} & \frac{1}{5} \end{bmatrix} \begin{bmatrix} \frac{1}{2} \\ \frac{1}{2} \\ 0 \end{bmatrix} = \begin{bmatrix} \frac{1}{8} \\ \frac{9}{20} \\ \frac{17}{20} \end{bmatrix}, \quad X_2 = \begin{bmatrix} 0 & \frac{1}{4} & \frac{3}{5} \\ \frac{2}{5} & \frac{1}{2} & \frac{1}{5} \\ \frac{3}{5} & \frac{1}{4} & \frac{1}{5} \end{bmatrix} \begin{bmatrix} \frac{1}{8} \\ \frac{9}{20} \\ \frac{17}{40} \end{bmatrix} = \begin{bmatrix} \frac{147}{400} \\ \frac{9}{25} \\ \frac{109}{400} \end{bmatrix} = \begin{bmatrix} .3675 \\ .36 \\ .2725 \end{bmatrix}$

7. This is an absorbing matrix since state 1 is an absorbing state and it is possible to go from any nonabsorbing state to state 1.

9. This is not an absorbing stochastic matrix since there is no absorbing matrix.

11. We solve the matrix equation

$$\begin{bmatrix} .6 & .4 \\ .3 & .7 \end{bmatrix} \begin{bmatrix} x \\ y \end{bmatrix} = \begin{bmatrix} x \\ y \end{bmatrix}$$

or equivalently, the system of equations

$-.4x + .3y = 0$

$.4x - .3y = 0$

$x + y = 1$

Solving this system of equations, we find the steady-state distribution vector to be

$\begin{bmatrix} \frac{4}{9} \\ \frac{5}{9} \end{bmatrix}$ and the steady-state matrix to be $\begin{bmatrix} \frac{4}{9} & \frac{4}{9} \\ \frac{5}{9} & \frac{5}{9} \end{bmatrix}$.

13. We solve the system

$$\begin{bmatrix} .6 & .4 & .3 \\ .2 & .2 & .2 \\ .2 & .4 & .5 \end{bmatrix} \begin{bmatrix} x \\ y \\ z \end{bmatrix} = \begin{bmatrix} x \\ y \\ z \end{bmatrix}$$

together with the equation $x + y + z = 1$, or equivalently, the system

$.6x + .4y + .3z = x$

$.2x + .2y + .2z = y$

$.2x + .4y + .5z = z$

$x + y + z = 1$

upon solving the system, we find the $x = .457$, $y = .20$, and $z = .343$,

and the steady-state distribution vector is given by $\begin{bmatrix} .457 \\ .20 \\ .343 \end{bmatrix}$ and the steady-state matrix

is $\begin{bmatrix} .457 & .457 & .457 \\ .20 & .20 & .20 \\ .343 & .343 & .343 \end{bmatrix}$.

15. a. The transition matrix for the Markov Chain is given by

$$T = \begin{matrix} & A & U & N \\ A & \begin{bmatrix} .85 & 0 & .10 \\ U & .10 & .95 & .05 \\ N & .10 & .05 & .85 \end{bmatrix} \end{matrix}$$

The probability vector describing the distribution of land 10 years ago is given by

$$\begin{bmatrix} .50 \\ .15 \\ .35 \end{bmatrix} \begin{matrix} A \\ U \\ N \end{matrix}.$$

To find the required probability vector, we compute

$$TX_0 = \begin{bmatrix} .85 & 0 & .10 \\ .10 & .95 & .05 \\ .05 & .05 & .85 \end{bmatrix} \begin{bmatrix} .50 \\ .15 \\ .35 \end{bmatrix} = \begin{bmatrix} .46 \\ .21 \\ .33 \end{bmatrix}$$

$$TX_1 = \begin{bmatrix} .85 & 0 & .10 \\ .10 & .95 & .05 \\ .05 & .05 & .85 \end{bmatrix} \begin{bmatrix} .46 \\ .21 \\ .33 \end{bmatrix} = \begin{bmatrix} .424 \\ .262 \\ .314 \end{bmatrix}.$$

Thus, the probability vector describing the distribution of land 10 years from now is

$$\begin{bmatrix} .424 \\ .262 \\ .314 \end{bmatrix}.$$

17. From

$$\begin{bmatrix} 1 & 2 \\ 3 & 5 \\ ④ & 6 \end{bmatrix}\begin{matrix} 1 \\ 3 \\ 4 \end{matrix} \leftarrow \text{ largest of the row minima}$$

$$\begin{matrix} 4 & 6 \end{matrix}$$
$$\uparrow \underline{\hspace{4cm}} \text{smaller of the column maxima}$$

we see that the game is strictly determined, and

a. the saddle point is 4.
b. the optimum strategy for the row player is to play row 3, whereas the optimum strategy for the column player is to play column 1.
c. the value of the game is 4.
d. the game favors the row player.

19. We first determine the largest of the row minima nd the smallest of the column maxima and display these elements as shown below.

$$\begin{bmatrix} ① & 3 & 6 \\ -2 & 4 & 3 \\ -5 & -4 & -2 \end{bmatrix}\begin{matrix} 1 \\ -2 \\ -5 \end{matrix} \leftarrow \text{ the largest of the row minima}$$

$$\begin{matrix} 1 & 4 & 6 \end{matrix}$$
$$\uparrow \underline{\hspace{4cm}} \text{the smallest of the column maxima}$$

The entry $a_{11} = 1$ is the saddle point of the game and we conclude that the game is strictly determined. The row player's optimal strategy is to play row 1 and the column player's optimal strategy is to play column 1. The value of the game is 1 and the game favors the row player.

21. We compute

$$E = PAQ = \begin{bmatrix} \frac{1}{2} & \frac{1}{2} \end{bmatrix}\begin{bmatrix} 4 & 8 \\ 6 & -12 \end{bmatrix}\begin{bmatrix} \frac{1}{4} \\ \frac{3}{4} \end{bmatrix} = -\frac{1}{4}.$$

23. We compute

$$E = PAQ = \begin{bmatrix} .2 & .4 & .4 \end{bmatrix} \begin{bmatrix} 3 & -1 & 2 \\ 1 & 2 & 4 \\ -2 & 3 & 6 \end{bmatrix} \begin{bmatrix} .2 \\ .6 \\ .2 \end{bmatrix}$$

$$= \begin{bmatrix} .2 & 1.8 & 4.4 \end{bmatrix} \begin{bmatrix} .2 \\ .6 \\ .2 \end{bmatrix} = 2.$$

The expected payoff for the game is 2.

25. The game under consideration has no saddle point and is accordingly nonstrictly determined. Using the formulas for determing the optimal mixed strategies for a 2×2 game with $a = 1$, $b = -2$, $c = 0$, and $d = 3$, we find that

$$p_1 = \frac{d-c}{a+d-b-c} = \frac{3-0}{1+3+2-0} = \frac{3}{6} = \frac{1}{2}$$

$$p_2 = 1 - p_1 = 1 - \frac{1}{2} = \frac{1}{2}$$

so that the row player's optimal mixed strategy is given by $P = \begin{bmatrix} \frac{1}{2} & \frac{1}{2} \end{bmatrix}$.
Next, we compute

$$q_1 = \frac{d-b}{a+d-b-c} = \frac{3+2}{1+3+2-0} = \frac{5}{6}$$

$$q_2 = 1 - q_1 = 1 - \frac{5}{6} = \frac{1}{6}.$$

Thus, the optimal strategy for the column player is given by

$$Q = \begin{bmatrix} \frac{5}{6} \\ \frac{1}{6} \end{bmatrix}.$$

To determine whether the game favors one player over the other, we compute the expected value of the game which is given by

$$E = \frac{ad - bc}{a+d-b-c} = \frac{(1)(3)-(-2)(0)}{1+3+2-0} = \frac{3}{6} = \frac{1}{2}$$

or 1/2 units for each play of the game. These results imply that the gme favors the row player.

27. Using the formulas for the optimal strategies in a 2×2 nonstrictly determined game, we see that the optimal mixed strategy for the row player is

$$P = \begin{bmatrix} \frac{1}{10} & \frac{9}{10} \end{bmatrix}$$

and the optimal mixed strategy for the column player is $Q = \begin{bmatrix} \frac{4}{5} \\ \frac{1}{5} \end{bmatrix}$. The value of the

game is $PAQ = [1.2]$ and the game favors the row player.

29. a. The required payoff matrix is given by

$$\begin{array}{cc} & \text{Record World} \\ & \begin{array}{cc} \$7 & \$8 \end{array} \\ \text{Disco-Mart} \begin{array}{c} \$7 \\ \$8 \end{array} & \begin{bmatrix} .5 & .7 \\ .4 & .5 \end{bmatrix} \begin{array}{c} .5 \\ .4 \end{array} \\ & \begin{array}{cc} .5 & .7 \end{array} \end{array}$$

   b.   Upon finding the larger of the row minima and the smaller of the column maxima, we see that the entry $a_{11} = .5$ is a saddle point, and , consequently, that the game is strictly determined. Thus, the optimal price that each company should sell the compact disc label for is $7.